NANOTECHNOLOGY SCIENCE AND TECHNOLOGY

NANOTECHNOLOGY: LOOKING AHEAD

NANOTECHNOLOGY SCIENCE AND TECHNOLOGY

Additional books in this series can be found on Nova's website under the Series tab.

Additional E-books in this series can be found on Nova's website under the E-books tab.

NANOTECHNOLOGY SCIENCE AND TECHNOLOGY

NANOTECHNOLOGY: LOOKING AHEAD

JULIAN S. MOSER
EDITOR

Nova Science Publishers, Inc.
New York

For permission to use material from this book please contact us:
Telephone 631-231-7269; Fax 631-231-8175
Web Site: http://www.novapublishers.com

LIBRARY OF CONGRESS CATALOGING-IN-PUBLICATION DATA
Nanotechnology : looking ahead / editor, Julian S. Moser.
p. cm.
Includes index.
ISBN 978-1-61761-785-0 (hardcover)
1. Nanotechnology--Forecasting--United States 2. Nanotechnology--Government policy--United States. I. Moser, Julian S. II. Title.
T174.7.N3747 2010
620'.50973--dc22
2010036422

Published by Nova Science Publishers, Inc. † New York

CONTENTS

PREFACE

Nanotechnology research and development is directed toward the understanding and control of matter at dimensions of roughly 1 to 100 nanometers. At this size, the physical, chemical, and biological properties of materials can differ in fundamental and potentially useful ways from the properties of individual atoms and molecules, or bulk matter. Congress continues to demonstrate interest in and support for nanotechnology due to what many believe is its extraordinary potential for delivering economic growth, high-wage jobs, and other societal benefits to the nation. This book provides an overview of the future of nanotechnology and the federal research and development investments under the National Nanotechnology Initiative; U.S. international competitiveness in nanotechnology; and environmental health, and safety concerns in the development of nanotechnology.

Chapter 1- Nanoscale science, engineering and technology—commonly referred to collectively as nanotechnology—is believed by many to offer extraordinary economic and societal benefits. Congress has demonstrated continuing support for nanotechnology and has directed its attention primarily to three topics that may affect the realization of this hoped for potential: federal research and development (R&D) in nanotechnology; U.S. competitiveness; and environmental, health, and safety (EHS) concerns. This chapter provides an overview of these topics—which are discussed in more detail in other CRS reports—and two others: nanomanufacturing and public understanding of and attitudes toward nanotechnology.

The development of this emerging field has been fostered by significant and sustained public investments in nanotechnology R&D. Nanotechnology R&D is directed toward the understanding and control of matter at dimensions of roughly 1 to 100 nanometers. At this size, the properties of matter can differ in fundamental and potentially useful ways from the properties of individual atoms and molecules and of bulk matter. Since the launch of the National Nanotechnology Initiative (NNI) in 2000 through FY2010, Congress has appropriated approximately $12.4 billion for nanotechnology R&D. In addition, the President requested an additional $1.8 billion in funding for nanotechnology R&D for FY20 11. More than 60 nations have established similar programs. In 2006 alone, total global public R&D investments reached an estimated $6.4 billion, complemented by an estimated private sector investment of $6.0 billion. Data on economic outputs used to assess competitiveness in mature technologies and industries, such as revenues and market share, are not available for assessing nanotechnology. Alternatively, data on inputs (e.g., R&D expenditures) and non-financial outputs (e.g., scientific papers, patents) may provide insight into the current U.S. position and serve as bellwethers of future competitiveness. By these criteria, the United

States appears to be the overall global leader in nanotechnology, though some believe the U.S. lead may not be as large as it was for previous emerging technologies.

Chapter 2- Nanotechnology—a term encompassing the science, engineering, and applications of submicron materials—involves the harnessing of unique physical, chemical, and biological properties of nanoscale substances in fundamentally new and useful ways. The economic and societal promise of nanotechnology has led to substantial and sustained investments by governments and companies around the world. In 2000, the United States launched the world's first national nanotechnology program. From FY2001 through FY2010, the federal government invested approximately $12.4 billion in nanoscale science, engineering, and technology through the U.S. National Nanotechnology Initiative (NNI). U.S. companies and state governments have invested billions more. President Obama has requested an additional $1.8 billion in NNI funding for FY2011. As a result of this focus and these investments, the United States has, in the view of many experts, emerged as a global leader in nanotechnology. However, the competition for global leadership in nanotechnology is intensifying as countries and companies around the world increase their investments.

Chapter 3- The National Nanotechnology Initiative (NNI) is a U.S. Government research and development (R&D) program of twenty-five agencies working together toward the common challenging vision of *"a future in which the ability to understand and control matter at the nanoscale leads to a revolution in technology and industry that benefits society."* The combined, coordinated efforts of these agencies have accelerated discovery, development, and deployment of nanotechnology towards agency missions and the broader national interest. Established in 2001,[1] the NNI involves nanotechnology-related activities by the 25 agencies shown in Table 1, 15 of which have budgets for nanotechnology R&D for 2011.

The National Nanotechnology Initiative is managed within the framework of the National Science and Technology Council (NSTC), the Cabinet-level council by which the President coordinates science and technology policy across the Federal Government. The Nanoscale Science, Engineering, and Technology (NSET) Subcommittee of the NSTC coordinates planning, budgeting, program implementation, and review of the initiative. The NSET Subcommittee is composed of representatives from agencies participating in the NNI. A listing of official NSET Subcommittee members is provided at the front of this chapter. Contact information for NSET Subcommittee participants is provided in Appendix B. The National Nanotechnology Coordination Office (NNCO) acts as the primary point of contact for information on the NNI; provides technical and administrative support to the NSET Subcommittee; supports the subcommittee in the preparation of multiagency planning, budget, and assessment documents, including this chapter; develops, updates, and maintains the NNI website (http://www.nano.gov); and provides public outreach on behalf of the NNI.

Chapter 4- The President's Council of Advisors on Science and Technology (PCAST) is an advisory group of the Nation's leading scientists and engineers, appointed by the President to augment the science and technology advice available to him from inside the White House and from cabinet departments and other Federal agencies. PCAST is consulted about and often makes policy recommendations concerning the full range of issues where understandings from the domains of science, technology, and innovation bear potentially on the policy choices before the President. PCAST is administered by the White House Office of Science and Technology Policy (OSTP).

Chapter 5- Enhanced abilities to understand and manipulate matter at the molecular and atomic levels promise a wave of significant new technologies over the next five decades.

Dramatic breakthroughs will occur in diverse areas such as medicine, communications, computing, energy, and robotics. These changes will generate large amounts of wealth and force wrenching changes in existing markets and institutions.

This chapter discusses the range of sciences currently covered by nanotechnology. It begins with a description of what nanotechnology is and how it relates to previous scientific advances. It then describes the most likely future development of different technologies in a variety of fields. The chapter also reviews the government's current nanotechnology policy and makes some suggestions for improvement.

In: Nanotechnology: Looking Ahead
Editor: Julian S. Moser

ISBN: 978-1-61761-785-0

Chapter 1

NANOTECHNOLOGY: A POLICY PRIMER

John F. Sargent Jr.

SUMMARY

Nanoscale science, engineering and technology—commonly referred to collectively as nanotechnology—is believed by many to offer extraordinary economic and societal benefits. Congress has demonstrated continuing support for nanotechnology and has directed its attention primarily to three topics that may affect the realization of this hoped for potential: federal research and development (R&D) in nanotechnology; U.S. competitiveness; and environmental, health, and safety (EHS) concerns. This chapter provides an overview of these topics—which are discussed in more detail in other CRS reports—and two others: nanomanufacturing and public understanding of and attitudes toward nanotechnology.

The development of this emerging field has been fostered by significant and sustained public investments in nanotechnology R&D. Nanotechnology R&D is directed toward the understanding and control of matter at dimensions of roughly 1 to 100 nanometers. At this size, the properties of matter can differ in fundamental and potentially useful ways from the properties of individual atoms and molecules and of bulk matter. Since the launch of the National Nanotechnology Initiative (NNI) in 2000 through FY2010, Congress has appropriated approximately $12.4 billion for nanotechnology R&D. In addition, the President requested an additional $1.8 billion in funding for nanotechnology R&D for FY20 11. More than 60 nations have established similar programs. In 2006 alone, total global public R&D investments reached an estimated $6.4 billion, complemented by an estimated private sector investment of $6.0 billion. Data on economic outputs used to assess competitiveness in mature technologies and industries, such as revenues and market share, are not available for assessing nanotechnology. Alternatively, data on inputs (e.g., R&D expenditures) and non-financial outputs (e.g., scientific papers, patents) may provide insight into the current U.S. position and serve as bellwethers of future competitiveness. By these criteria, the United States appears to be the overall global leader in nanotechnology, though some believe the U.S. lead may not be as large as it was for previous emerging technologies.

Some research has raised concerns about the safety of nanoscale materials. There is general agreement that more information on EHS implications is needed to protect the public and the environment; to assess and manage risks; and to create a regulatory environment that fosters prudent investment in nanotechnology-related innovation. Nanomanufacturing—the bridge between nanoscience and nanotechnology products—may require the development of new technologies, tools, instruments, measurement science, and standards to enable safe, effective, and affordable commercial-scale production of nanotechnology products. Public understanding and attitudes may also affect the environment for R&D, regulation, and market acceptance of products incorporating nanotechnology.

In 2003, Congress enacted the 21st Century Nanotechnology Research and Development Act providing a legislative foundation for some of the activities of the NNI, addressing concerns, establishing programs, assigning agency responsibilities, and setting authorization levels. Legislation has been introduced in the House (H.R. 554) and Senate (S. 1482) that would amend the act. The House passed H.R. 554 on February 11, 2009; the Senate has not acted on this legislation. On May 7, 2010, the House Committee on Science and Technology reported the America COMPETES Reauthorization Act of 2010 (H.R. 5116) which includes, as Title I, Subtitle A, the National Nanotechnology Initiative Amendments Act of 2010. Provisions of this subtitle are nearly identical to the provisions of H.R. 554. On May 13, the House voted to recommit the bill to committee with instructions. Subsequently, the bill was pulled from consideration at the request of the chairman of the House Committee on Science and Technology.

Overview

Congress continues to demonstrate interest in and support for nanotechnology due to what many believe is its extraordinary potential for delivering economic growth, high-wage jobs, and other societal benefits to the nation. To date, the Science and Technology Committee in the House and Senate Committee on Commerce, Science, and Transportation have directed their attention primarily to three topics that may affect the United States' realization of this hoped for potential: federal research and development (R&D) investments under the National Nanotechnology Initiative (NNI); U.S. international competitiveness; and environmental, health, and safety (EHS) concerns. This chapter provides a brief overview of these topics—which are discussed in greater detail in CRS reports [1]—and two other subjects of interest to Congress: nanomanufacturing and public attitudes toward, and understanding of, nanotechnology.

Nanotechnology research and development is directed toward the understanding and control of matter at dimensions of roughly 1 to 100 nanometers. At this size, the physical, chemical, and biological properties of materials can differ in fundamental and potentially useful ways from the properties of individual atoms and molecules, on the one hand, or bulk matter, on the other hand.

In 2000, President Clinton launched the NNI to coordinate federal R&D efforts and promote U.S. competitiveness in nanotechnology. Congress first funded the NNI in FY2001 and has provided increased appropriations for nanotechnology R&D in each subsequent year. In 2003, Congress enacted the 21st Century Nanotechnology Research and Development Act

(P.L. 108-153). The act provided a statutory foundation for the NNI, established programs, assigned agency responsibilities, authorized funding levels, and initiated research to address key issues.

Federal R&D investments are focused on advancing understanding of fundamental nanoscale phenomena and on developing nanomaterials, nanoscale devices and systems, instrumentation, standards, measurement science, and the tools and processes needed for nanomanufacturing. NNI appropriations also fund the construction and operation of major research facilities and the acquisition of instrumentation. Finally, the NNI supports research directed at identifying and managing potential environmental, health, and safety impacts of nanotechnology, as well as its ethical, legal and societal implications.

Most current applications of nanotechnology are evolutionary in nature, offering incremental improvements in existing products and generally modest economic and societal benefits. For example, nanotechnology is being used: in automobile bumpers, cargo beds, and step-assists to reduce weight, increase resistance to dents and scratches, and eliminate rust; in clothes to increase stain- and wrinkle-resistance; and in sporting goods, such as baseball bats and golf clubs, to improve performance.

In the longer term, nanotechnology may deliver revolutionary advances with profound economic and societal implications. Potential applications discussed by the technology's proponents involve various degrees of speculation and varying time-frames. The examples below suggest areas where such possible revolutionary advances may emerge, and early research and development efforts that may provide insights into how such advances may be achieved.

- **Detection and treatment technologies for cancer and other deadly diseases.** Current nanotechnology disease detection efforts include the development of sensors that can identify biomarkers, such as altered genes, that may provide an early indicator of cancer. One approach uses carbon nanotubes and nanowires to identify the unique molecular signals of cancer biomarkers. Another approach uses nanoscale cantilevers—resembling a row of diving boards—treated with molecules that bind only with cancer biomarkers. When these molecules bind, the additional weight bends the cantilevers indicating the presence and concentration of these biomarkers. Nanotechnology holds promise for showing the presence, location, and/or contours of cancer, cardiovascular disease, or neurological disease. Current R&D efforts employ metallic, magnetic, and polymeric nanoparticles with strong imaging characteristics attached to an antibody or other agent that binds selectively with targeted cells. The imaging results can be used to guide surgical procedures and to monitor the effectiveness of non-surgical therapies in killing the disease or slowing its growth. Nanotechnology may also offer new cancer treatment approaches. For example, nanoshells with a core of silica and an outer metallic shell can be engineered to concentrate at cancer lesion sites. Once at the sites, a harmless energy source (such as near-infrared light) can be used to cause the nanoshells to heat, killing the cancer cells they are attached to. Another treatment approach targets delivery of tiny amounts of a chemotherapy drug to cancer cells. In this approach the drug is encapsulated inside a nanoshell that is engineered to bind with an antigen on the cancer cell. Once bound, the nanoshell dissolves, releasing the chemotherapy drug, killing the cancer cell. Such a targeted delivery approach could reduce the amount of chemotherapy drug needed to kill the cancer cells, reducing the side effects of chemotherapy.[2]

- **Clean, inexpensive, renewable power through energy creation, storage, and transmission technologies.** Nanoscale semiconductor catalysts and additives show promise for improving the production of hydrogen from water using sunlight. The optical properties of these nanoscale catalysts allow the process to use a wider spectrum of sunlight. Similarly, nanostructured photovoltaic devices (e.g., solar panels) may improve the efficiency of converting sunlight into electricity by using a wider spectrum of sunlight. Improved hydrogen storage, a key challenge in fuel cell applications, may be achieved by tapping the chemical properties and large surface area of certain nanostructured materials. In addition, carbon nanotube fibers have the potential for reducing energy transmission losses from approximately 7% (using copper wires) to 6% (using carbon nanotube fibers), an equivalent annual energy savings in the United States of 24 million barrels of oil.[3]
- **Universal access to clean water.** Nanotechnology water desalination and filtration systems may offer affordable, scalable, and portable water filtration systems. Filters employing nanoscale pores work by allowing water molecules to pass through, but prevent larger molecules, such as salt ions and other impurities (e.g., bacteria, viruses, heavy metals, and organic material), from doing so. Some nanoscale filtration systems also employ a matrix of polymers and nanoparticles that serve to attract water molecules to the filter and to repel contaminants.[4]
- **High-density memory devices.** A variety of nanotechnology applications may hold the potential for improving the density of memory storage. For example, IBM has demonstrated the potential to create high-density memory devices (with an estimated storage capacity of 1 terabyte per square inch) by storing information mechanically using thermal-mechanical nanoscale probes to punch nanoscale indentations into a thin plastic film. The probes can be used to read and write data in parallel.[5]
- **Higher crop yield and improved nutrition.** Higher crop yield might be achieved using nanoscale sensors that detect the presence of a virus or disease- infecting particle. Early, location-specific detection may allow for rapid and targeted treatment of affected areas, increasing yield by preventing losses.[6] Nanotechnology also offers the potential for improved nutrition. Some companies are exploring the development of nanocapsules that release nutrients targeted at specific parts of the body at specific times.[7]
- **Self-healing materials.** Nanotechnology may offer approaches that enable materials to "self-heal" by incorporating, for example, nanocontainers of a repair substance (e.g., an epoxy) throughout the material. When a crack or corrosion reaches a nanocontainer, the nanocontainer could be designed to open and release its repair material to fill the gap and seal the crack.[8]
- **Sensors that can warn of minute levels of toxins and pathogens in air, soil or water.** Microfluidic and nanocantilever sensors (discussed earlier) may be engineered to detect specific pathogens (e.g., bacteria, virus) or toxins (e.g., sarin gas, hydrogen cyanide) by detecting their unique molecular signals or through selective binding with an engineered nanoparticle.
- **Environmental remediation of contaminated sites.** The high surface-to-volume ratio, high reactivity, and small size of some nanoscale particles (e.g., nanoscale iron) may offer more effective and less costly solutions to environmental contamination. By injecting engineered nanoparticles into the ground, these characteristics can be employed to enable

the particles to move more easily through a contaminated site and bond more readily with targeted contaminants.[9]

Nanotechnology is also expected by some to make substantial contributions to federal missions such as national defense, homeland security, and space exploration and commercialization.

U.S. private sector nanotechnology R&D is now estimated to be twice that of public funding. In general, the private sector's efforts are focused on translating fundamental knowledge and prototypes into commercial products; developing new applications incorporating nanoscale materials; and developing technologies, methods, and systems for commercial-scale manufacturing.

Many other nations and firms around the world are also making substantial investments in nanotechnology to reap its potential benefits. With so much potentially at stake, some members of Congress have expressed interest and concerns about the U.S. competitive position in nanotechnology R&D and success in translating R&D results to commercial products. This has led to an increased focus on potential barriers to commercialization efforts, including the readiness of technologies, systems, and processes for large-scale nanotechnology manufacturing; potential environmental, health, and safety (EHS) effects of nanoscale materials; public understanding and attitudes toward nanotechnology; and other related issues.

While many provisions of the 21st Century Nanotechnology Research and Development Act have no sunset provision, FY2008 was the last year of agency authorizations included in the act. Legislation to amend and reauthorize the act was introduced in the House (H.R. 5940, 110th Congress) and the Senate (S. 3274, 110th Congress) in the 110th Congress. Both bills were titled the National Nanotechnology Initiative Amendments Act of 2008. The House passed H.R. 5940 by a vote of 407-6; the Senate did not act on S. 3274. In January 2009, H.R. 554, the National Nanotechnology Initiative Amendments Act of 2009, was introduced in the 111th Congress. The act contains essentially the same provisions as H.R. 5940 (1 10th Congress). In February 2009, the House passed the bill by voice vote under a suspension of the rules. The bill was referred to the Senate Committee on Commerce, Science, and Transportation; no further action has been taken on this bill.

On May 7, 2010, the House Committee on Science and Technology reported the America COMPETES Reauthorization Act of 2010 (H.R. 5116; H.Rept. 111 -478)[10] which includes, as Title I, Subtitle A, the National Nanotechnology Initiative Amendments Act of 2010. The provisions of this subtitle are nearly identical to the provisions of H.R. 554. On May 13, the House voted to recommit the bill to committee with instructions. Subsequently, the bill was pulled from consideration at the request of the chairman of the House Committee on Science and Technology.

This chapter provides a macro-level view of federal R&D investments, U.S. competitiveness in nanotechnology, and EHS-related issues.

Table 1. NNI Funding, by Agency (in Millions of Current Dollars)

Agency	FY 2001 Actual	FY 2002 Actual	FY 2003 Actual	FY 2004 Actual	FY 2005 Actual	FY 2006 Actual	FY 2007 Actual	FY 2008 Actual	FY 2009 Actual	FY2009 ARRA	FY2010 Estimate	FY2011 Request
Department of Defense[a]	125	224	220	291	352	424	450	460	459		436	349
National Science Foundation	150	204	221	256	335	360	389	409	409	101	418	401
Department of Energy	88	89	134	202	208	231	236	245	333	293	373	424
National Institutes of Health (DHHS)[b]	40	59	78	106	165	192	215	305	343	73	361	382
National Institute of Standards and Technology (DOC)	33	77	64	77	79	78	88	86	93	43	114	108
National Aeronautics and Space Administration	22	35	36	47	45	50	20	17	14		14	16
Environmental Protection Agency	5	6	5	5	7	5	8	12	12		18	20
Other Agencies	1	3	2	5	9	13	19	22	40		48	62
TOTAL[c]	464	697	760	989	1,200	1,351	1,425	1,554	1,702	511	1,782	1,762

Source: NNI website. http://www.nano.gov/html/about/funding.html

a. According to NSTC, the Department of Defense budgets for FY2006-FY20 10 include congressionally directed funding outside the NNI plan. The extent to which such funding is included or not included in reporting of funding in earlier fiscal years is uncertain.

b. According to NIH, the agency has adopted the Research, Condition, and Disease Categorization (RCDC) system to provide more consistent and transparent information to the public about NIH research. The shift to the RCDC process of categorization changes the way individual research projects are assigned to categories. This change will result in some differences in total dollar amounts between the 2008 reports and those issued in previous years. Any difference, whether an increase or decrease in funding levels, does not necessarily reflect a change in the amount of money the NIH received from Congress or a change in the actual content of the NIH research portfolio. For more information, please go to: http://report.nih.gov/rcdc/reasons /default.aspx.

c. Numbers may not add due to rounding of agency budget figures.

The National Nanotechnology Initiative

President Clinton launched the National Nanotechnology Initiative in 2000, establishing a multi- agency program to coordinate and expand federal efforts to advance the state of nanoscale science, engineering, and technology, and to position the United States to lead the world in its development and commercialization. The NNI is comprised of 13 federal agencies that receive appropriations to conduct and fund nanotechnology R&D and 12 other federal agencies with responsibilities for health, safety, and environmental regulation; trade; education; training; intellectual property; international relations; and other areas that might affect nanotechnology. EPA both conducts R&D and has regulatory responsibilities.

Congress has played a central role in the NNI, providing appropriations for the conduct of nanotechnology R&D (discussed below), establishing programs, and creating a legislative foundation for some of the activities of the NNI through enactment of the 21st Century Nanotechnology Research and Development Act of 2003. The act authorized appropriations FY2005 through FY2008 for five NNI agencies–the National Science Foundation (NSF), Department of Energy (DOE), National Aeronautics and Space Administration (NASA), Department of Commerce (DOC) National Institute of Standards and Technology (NIST), and Environmental Protection Agency (EPA). Congress remains actively engaged in the NNI. Legislation has been introduced in the House (National Nanotechnology Initiative Amendments Act of 2009, H.R. 554) and Senate (National Nanotechnology Initiative Amendments Act of 2009, S. 1482) that would amend the act. The House passed H.R. 554 on February 11, 2009. The Senate has not acted on this legislation. The 111th Congress may address policy issues related to the NNI through this or other legislation.

Structure

The NNI is coordinated within the White House through the National Science and Technology Council (NSTC) Nanoscale Science, Engineering, and Technology (NSET) subcommittee. The NSET subcommittee is comprised of representatives from 25 federal agencies, White House Office of Science and Technology Policy (OSTP) and Office of Management and Budget.[11] The NSET subcommittee has established several working groups, including the National Environmental and Health Implications (NEHI), National Innovation and Liaison with Industry (NILI), Global Issues in Nanotechnology (GIN), Nanomanufacturing, and Nanotechnology Public Engagement and Communications (NPEC) working groups. The National Nanotechnology Coordination Office (NNCO) provides administrative and technical support to the NSET subcommittee.

Funding

Funding for the NNI is provided through appropriations to each of the NNI-participating agencies. The NNI has no centralized funding. Overall NNI funding is calculated by aggregating the nanotechnology-related expenditures of each NNI agency. Funding remains concentrated in the original six NNI agencies[12] which accounts for 96.3% of NNI funding in

FY2010. The NNI funds fundamental and applied nanotechnology R&D, multidisciplinary centers of excellence, and key research infrastructure. It also supports efforts to address societal implications of nanotechnology, including ethical, legal, EHS, and workforce issues.

For FY20 10, Congress appropriated an estimated $1.8 billion for nanotechnology R&D, nearly four times the $464 million appropriated for nanotechnology R&D in 2001. In total, Congress has appropriated approximately $12.4 billion for the NNI since FY2001. President Obama has requested $1.76 1 billion for nanotechnology R&D in FY2010, a $20 million (1.1%) decrease below the estimated FY2010 funding level of $1.781 billion. The chronology of NNI funding is detailed in Table 1.

SELECTED ISSUES

U.S. Competitiveness

Nanotechnology is largely still in its infancy. Accordingly, measures such as revenues, market share, and global trade statistics—which are often used to assess and track U.S. competitiveness in other more mature technologies and industries—are not available for assessing the relative U.S. position internationally in nanotechnology. To date, the federal government does not collect data on nanotechnology-related revenues, trade or employment, nor is comparable international government data available. Nevertheless, many experts believe that the United States is the global leader in nanotechnology. However, some of these experts believe that in contrast to many previous emerging technologies—such as semiconductors, satellites, software, and biotechnology—the U.S. lead is narrower, and the investment level, scientific and industrial infrastructure, technical capabilities, and science and engineering workforces of other nations are more substantial than in the past.

In the absence of comprehensive and reliable economic output data (e.g., revenues, market share, trade), indicators such as inputs (e.g., public and private research investments) and non-financial outputs (e.g., scientific papers, patents) have been used to gauge a nation's competitive position in emerging technologies. By these measures (discussed below), the United States appears to lead the world, generally, in nanotechnology. However, R&D investments, scientific papers, and patents may not provide reliable indicators of the United States' current or future competitive position. Scientific and technological leadership may not necessarily result in commercial leadership or national competitiveness for a variety of reasons:

- Basic research in nanotechnology may not translate into viable commercial applications.
- Basic research is generally available to all competitors.
- U.S.-based companies may conduct production and other work outside of the United States.
- U.S.-educated foreign students may return home to conduct research and create new businesses.
- U.S. companies with leading-edge nanotechnology capabilities and/or intellectual property may be acquired by foreign competitors.

- U.S. policies or other factors may prohibit nanotechnology commercialization, make it unaffordable, or make it less attractive than foreign alternatives.
- Aggregate national data may be misleading as countries may establish global leadership in niche areas of nantoechnology.

With these caveats, the following section reviews input and non-economic output measures as indicators of the U.S. competitive position in nanotechnology.

Global Funding

The United States has led, and continues to lead, all nations in known public investments in nanotechnology R&D, though the estimated U.S. share of global public investments has fallen as other nations have established similar programs and increased funding. Using a currency exchange rate comparison, the United States ranks ahead of all other nations, with federal and state investments of $1.78 billion in 2006 (27.8% of global public R&D investments), followed by Japan ($975 million, 15.2%) and Germany ($563 million, 8.8%). When national investments are adjusted using purchasing power parity (PPP) exchange rates,[13] the United States remains the world leader, but China ranked second in public nanotechnology spending in 2005.[14]

Private investments in nanotechnology R&D come from two primary sources, corporations and venture capital investors. On a PPP comparison basis, the United States led the world in 2006 in corporate R&D investments in nanotechnology with an estimated $1.9 billion investment, followed by Japan with $1.7 billion. In total, U.S. and Japan-based companies accounted for nearly three-fourths of global corporate investment in nanotechnology R&D in 2006. China ranks fifth in corporate investment, accounting for approximately 3% of global private R&D investments.[15] Lux Research, an emerging technologies consulting firm, estimates that global nanotechnology venture capital investment in 2007 was $702 million, of which $632 million went to U.S.-based firms.[16]

Scientific Papers

The quantity of peer-reviewed scientific papers is considered by some to be an indicator of a nation's scientific leadership. A study by the National Bureau of Economic Research in 2005 reported that the U.S. share was a world-leading 24%, but that this represented a decline from approximately 40% in the early 1 990s, concluding:

> Taken as a whole these data confirm that the strength and depth of the American science base points to the United States being the dominant player in nanotechnology for some time to come, while the United States also faces significant and increasing international competition.[17]

One measure of the importance of a scientific paper is the number of times it is cited in other papers. An analysis by Evaluametrics, Ltd. reports that nanotechnology papers attributed to the United States are much more frequently cited than those attributed to China, the nations of the European Union (EU27), and the rest of the world as a whole. This held true overall and separately in each of the four disciplines examined (biology, chemistry, engineering, and physics). The U.S. lead was particularly pronounced in biology. China fell below the world average number of citations in each discipline, as well as overall. The EU27 performed near the world average in engineering and physics, and somewhat higher in chemistry.

Patents

Patent counts—assessments of how many patents are issued to individuals or institutions of a particular country—are frequently used to assess technological competitiveness. By this measure, the U.S. competitive position appears to be strong. A 2007 U.S. Patent and Trademark Office analysis of patents in the United States and in other nations stated that U.S.-origin inventors and assignees/owners have:

- the most nanotechnology-related U.S. patents by a wide margin;
- the most nanotechnology-related patent publications globally, but by a narrower margin (followed closely by Japan); and
- the most nanotechnology-related inventions that have patent publications in three or more countries, 31.7%, followed by Japan (26.9%), Germany (11.3%), Korea (6.6%), and France (3.6%).[18]

ENVIRONMENTAL, HEALTH, AND SAFETY IMPLICATIONS

Key policy issues associated with U.S. competitiveness in nanotechnology include environmental, health, and safety (EHS) concerns, nanomanufacturing, and public understanding and attitudes. EHS concerns include both direct consequences for health, safety, and the environment, and how uncertainty about EHS implications and potential regulatory responses might affect U.S. competitiveness. One such effect might be the discouragement of investment in nanotechnology due to the possibility of regulations that might bar products from the market, impose high regulatory compliance costs, or result in product liability claims and clean-up costs.

Some of the unique properties of nanoscale materials—e.g., small size, high surface area-to-volume ratio—have given rise to concerns about their potential implications for health, safety, and the environment. While nanoscale particles occur naturally and as incidental by-products of other human activities (e.g., soot),[19] EHS concerns have been focused primarily on nanoscale materials that are intentionally engineered and produced.

Much of the public dialogue about risks associated with nanotechnology has focused on carbon nanotubes (CNTs) and other fullerenes (molecules formed entirely of carbon atoms in the form of a hollow sphere, ellipsoid, or tube) since they are currently being manufactured and are among the most promising nanomaterials. These concerns have been amplified by some research on the effects of CNTs on animals, and on animal and human cells. For example, researchers have reported that carbon nanotubes inhaled by mice can cause lung tissue damage;[20] that buckyballs (spherical fullerines) caused brain damage in fish;[21] and that buckyballs can accumulate within cells and potentially cause DNA damage.[22] On the other hand, some research has found CNTs and fullerenes to be non-toxic. In addition, work at Rice University's Center for Biological and Environmental Nanotechnology conducted in 2005 found cell toxicity of CNTs to be low and that toxicity can be reduced further through simple chemical changes to the CNT 's surface.[23]

Among the potential EHS benefits of nanotechnology are applications that may reduce energy consumption, pollution, and greenhouse gas emissions; remediate environmental damage; cure, manage, or prevent deadly diseases; and offer new materials that protect against impacts, self- repair to prevent catastrophic failure, or change in ways that provide protection and medical aid to soldiers on the battlefield.

Potential EHS risks of nanoscale particles in humans and animals depend in part on their potential to accumulate, especially in vital organs such as the lungs and brain, that might harm or kill, and diffusion in the environment that might harm ecosystems. For example, several products on the market today contain nanoscale silver, an effective antibacterial agent. Some scientists have raised concerns that the dispersion of nanoscale silver in the environment could kill microbes that are vital to ecosystems.

Like nanoscale silver, other nanoscale particles might produce both positive and negative effects. For example, some nanoscale particles have the potential to penetrate the blood-brain barrier, a structure that protects the brain from harmful substances in the blood. Currently, the barrier hinders the delivery of therapeutic agents to the brain.[24] The characteristics of some nanoscale materials may allow pharmaceuticals to be developed to purposefully and beneficially cross the blood-brain barrier and deliver medicine directly to the brain to treat, for example, a brain tumor. Alternatively, other nanoscale particles might unintentionally pass through this barrier and harm humans and animals.

There is widespread uncertainty about the potential EHS implications of nanotechnology. A survey of business leaders in the field of nanotechnology indicated that nearly two-thirds believe that "the risks to the public, the workforce, and the environment due to exposure to nano particles are 'not known,'" and 97% believe that it is very or somewhat important for the government to address potential health effects and environmental risks that may be associated with nanotechnology.[25]

Many stakeholders believe that concerns about potential detrimental effects of nanoscale materials and products on health, safety, and the environment—both real and perceived—must be addressed for a variety of reasons, including:

- protecting and improving human health, safety, and the environment;
- enabling accurate and efficient risk assessments, risk management, and cost-benefit trade-offs;
- creating a predictable, stable, and efficient regulatory environment that fosters investment in nanotechnology-related innovation;
- ensuring public confidence in the safety of nanotechnology research, engineering, manufacturing, and use;
- preventing the negative consequences of a problem in one application area of nanotechnology from harming the use of nanotechnology in other applications due to public fears, political interventions, or an overly-broad regulatory response; and

- ensuring that society can enjoy the widespread economic and societal benefits that nanotechnology may offer.

Policy issues associated with EHS impacts of nanotechnology include magnitude, timing, foci, and management of the federal investment in EHS research; adequacy of the current regulatory structures to protect public health and the environment; and cooperation with other nations engaged in nanotechnology R&D to ensure all are doing so in a responsible manner.

Nanomanufacturing

Securing the economic benefits and societal promise of nanotechnology requires the ability to translate knowledge of nanoscience into market-ready nanotechnology products. Nanomanufacturing is the bridge connecting nanoscience and nanotechnology products. Although some nanotechnology products have already entered the market, these materials and devices have tended to require only incremental changes in manufacturing processes. Generally, they are produced in a laboratory environment in limited quantities with a high-degree of labor intensity, high variability, and high costs. To make their way into safe, reliable, effective, and affordable commercial-scale production in a factory environment may require the development of new and unique technologies, tools, instruments, measurement science, and standards for nanomanufacturing.

Public Attitudes and Understanding

What the American people know about nanotechnology and the attitudes that they have toward it may affect the environment for research and development (especially support for public R&D funding), regulation, market acceptance of products incorporating nanotechnology, and, perhaps, the ability of nanotechnology to weather a negative event such as an accident or spill.

In 2007, the Woodrow Wilson International Center for Scholars' Project on Emerging Nanotechnologies (PEN) reported results of a nationwide poll of adults that found more than 42% had "heard nothing at all" about nanotechnology, while only 6% said they had "heard a lot." In addition, more than half of those surveyed felt they could not assess the relative value of nanotechnology's risks and benefits. Among those most likely to believe that benefits outweigh risks were those earning more than $75,000 per year, men, people who had previously heard "some" or "a lot" about nanotechnology, and those between the ages of 35 and 64. Alternatively, among those most likely to believe that the risks of nanotechnology outweigh benefits include people earning $30,000 or less; those with a high school diploma or less; women; racial and ethnic minorities; and those between the ages of 18 and 34 or over age 65.[26]

The PEN survey found a strong positive correlation between familiarity with and awareness of nanotechnology and perceptions that benefits will outweigh risks. However, the survey data also indicate that communicating with the public about nanotechnology in the absence of clear, definitive answers to EHS questions could create a higher level of uncertainty, discomfort, and opposition.

Congress expressed its belief in the importance of public engagement in the 21st Century Nanotechnology Research and Development Act of 2003 (15 U.S.C. § §7501 et seq.). The act calls for public input and outreach to be integrated into the NNI's efforts. The NNI has sought to foster public understanding through a variety of mechanisms, including written products, speaking engagements, a web-based information portal (nano.gov), informal education, and efforts to establish dialogues with key stakeholders and the general public. In addition, the NSET subcommittee has established a Nanotechnology Public Engagement and Communications working group to develop approaches by which the NNI can communicate more effectively with the public.

End Notes

1 For additional information on these issues, see CRS Report RL34401, *The National Nanotechnology Initiative: Overview, Reauthorization, and Appropriations Issues*, CRS Report RL34493, *Nanotechnology and U.S. Competitiveness: Issues and Options*, and CRS Report RL34614, *Nanotechnology and Environmental, Health, and Safety: Issues for Consideration*, all by John F. Sargent, and CRS Report RL34332, *Engineered Nanoscale Materials and Derivative Products: Regulatory Challenges*, by Linda-Jo Schierow.

2 National Cancer Institute website. http://nano.cancer.gov/resource_center/tech_backgrounder.asp.

3 *Nanoscience Research for Energy Needs*, Nanoscale Science, Engineering, and Technology Subcommittee, National Science and Technology Council, The White House, December 2004.

4 Abraham, M. "Today's Seawater is Tomorrow's Drinking Water," University of California at Los Angeles, November 6, 2006

5 Vettiger, P. "The 'millipede' - nanotechnology entering data storage," IEEE Transactions on Nanotechnology, March 2002. Vol 1. Issue 1. pp 29-55.

6 *Nanoscale Science and Engineering for Agriculture and Food Systems*, draft report on the National Planning Workshop, submitted to the Cooperative State Research, Education, and Extension Service of the U.S. Department of Agriculture, July 2003.

7 Wolfe, Josh. "Safer and Guilt-Free Nano Foods," Forbes.com, August 10, 2005.

8 Berger, Michael. "Nanomaterial heal thyself," Nanowerk Spotlight, June 13, 2007.

9 EPA website. http://es.epa.gov/ncer/nano/research/nano_remediation.html.

10 For additional information on this bill and related issues, see CRS Report R4123 1, *America COMPETES Reauthorization Act of 2010 and the America COMPETES Act: Selected Policy Issues*, coordinated by Heather B. Gonzalez.

11 NSET subcommittee members include Bureau of Industry and Security, DOC; Consumer Product Safety Commission; Cooperative State Research, Education, and Extension Service, Department of Agriculture (USDA); Department of Defense (DOD); Department of Education; DOE; Department of Homeland Security; Department of Justice; Department of Labor; Department of State; Department of Transportation; Department of the Treasury; EPA; Food and Drug Administration; Forest Service, USDA; Intelligence Technology Innovation Center; International Trade Commission; NASA; National Institutes of Health (NIH), Department of Health and Human Services (DHHS); National Institute for Occupational Safety and Health, Centers for Disease Control and Prevention, (DHHS); NIST, DOC; NSF; Nuclear Regulatory Commission; U.S. Geological Survey, Department of the Interior; and U.S. Patent and Trademark Office, DOC.

12 The original six agencies were the NSF, DOD, DOE, NIST, NASA, and NIH.

13 Purchasing power parity exchange rates seek to equalize the purchasing power of currencies in different countries for a given basket of goods and/or services.

14 *Profiting from International Nanotechnology*, Lux Research, Inc., December 2006.

15 *Profiting from International Nanotechnology*, Lux Research, Inc., December 2006.

16 Personal communication between CRS and Lux Research, April 28, 2008.

17 Zucker, L.G. and M.R. Darby. "Socio-Economic Impact of Nanoscale Science: Initial Results and Nanobank," National Bureau of Economic Research, March 2005.

18 Eloshway, Charles. "Nanotechnology Related Issues at the U.S. Patent and Trademark Office," Workshop on Intellectual Property Rights in Nanotechnology: Lessons from Experiences Worldwide, Brussels, Belgium, April 2007.

[19] Some naturally occurring nanoparticles cause adverse health effects. Studies on the effects of naturally occurring particles are numerous and inform R&D on engineered nanoparticles.

[20] Lam, C.; James, J.T.; McCluskey, R.; and Hunter, R. "Pulmonary toxicity of single-wall carbon nanotubes in mice 7 and 90 days after intratracheal instillation," *Toxicological Sciences*, September 2003. Vol 77. No. 1. pp 126-134.

[21] Oberdörster, Eva. "Manufactured Nanomaterials (Fullerenes, C60) Induce Oxidative Stress in the Brain of Juvenile Largemouth Bass," *Environmental Health Perspectives*, July 2004. Vol. 112. No. 10.

[22] "Understanding Potential Toxic Effects of Carbon-Based Nanomaterials," *Nanotech News*, National Cancer Institute Alliance for Nanotechnology in Cancer, July 10, 2006.

[23] "Modifications render carbon nanotubes nontoxic," press release, Rice University, October 2005.

[24] "Blood-Brain Barrier Breached by New Therapeutic Strategy," press release, National Institutes of Health, June 2007.

[25] "Survey of U.S. Nanotechnology Executives," Small Times Magazine and the Center for Economic and Civic Opinion at the University of Massachusetts-Lowell, Fall 2006.

[26] "Awareness of and Attitudes Toward Nanotechnology and Federal Regulatory Agencies: A Report of Findings," survey by Peter D. Hart Research Associates, Inc., for the Project on Emerging Nanotechnologies, September 2007.

In: Nanotechnology: Looking Ahead
Editor: Julian S. Moser
ISBN: 978-1-61761-785-0

Chapter 2

THE NATIONAL NANOTECHNOLOGY INITITATIVE: OVERVIEW, REAUTHORIZATION, AND APPROPRIATIONS ISSUES

John F. Sargent Jr.

SUMMARY

Nanotechnology—a term encompassing the science, engineering, and applications of submicron materials—involves the harnessing of unique physical, chemical, and biological properties of nanoscale substances in fundamentally new and useful ways. The economic and societal promise of nanotechnology has led to substantial and sustained investments by governments and companies around the world. In 2000, the United States launched the world's first national nanotechnology program. From FY2001 through FY2010, the federal government invested approximately $12.4 billion in nanoscale science, engineering, and technology through the U.S. National Nanotechnology Initiative (NNI). U.S. companies and state governments have invested billions more. President Obama has requested an additional $1.8 billion in NNI funding for FY2011. As a result of this focus and these investments, the United States has, in the view of many experts, emerged as a global leader in nanotechnology. However, the competition for global leadership in nanotechnology is intensifying as countries and companies around the world increase their investments.

Nanotechnology's complexity and intricacies, early stage of development (with commercial payoff possibly years away for many potential applications), and broad scope of potential applications engender a wide range of public policy issues. Maintaining U.S. technological and commercial leadership in nanotechnology poses a variety of technical and policy challenges, including development of technologies that will enable commercial scale manufacturing of nanotechnology materials and products; environmental, health, and safety (EHS) concerns; and maintenance of public confidence in its safety.

Congress established programs, assigned responsibilities, and initiated research and development (R&D) related to these issues in the 21st Century Nanotechnology Research and Development Act of 2003 (P.L. 108-153). While many provisions of this act have no sunset provision, FY2008 was the last year of agency authorizations included in the act. Legislation to amend and reauthorize the act was introduced in the House (H.R. 5940, 110th Congress) and the Senate (S. 3274, 110th Congress) in the 110th Congress. Both bills were titled the National Nanotechnology Initiative Amendments Act of 2008. The House passed H.R. 5940 by a vote of 407-6; the Senate did not act on S. 3274. In January 2009, H.R. 554, the National Nanotechnology Initiative Amendments Act of 2009, was introduced in the 111th Congress. The act contains essentially the same provisions as H.R. 5940 (110th Congress). In February 2009, the House passed the bill by voice vote under a suspension of the rules. The bill was referred to the Senate Committee on Commerce, Science, and Transportation; no further action has been taken. On May 7, 2010, the House Committee on Science and Technology reported the America COMPETES Reauthorization Act of 2010 (H.R. 5116) which includes, as Title I, Subtitle A, the National Nanotechnology Initiative Amendments Act of 2010. Provisions of this subtitle are nearly identical to the provisions of H.R. 554.

Proponents of the NNI assert that nanotechnology is one of the most important emerging and enabling technologies and that U.S. competitiveness, technological leadership, national security, and societal interests require an aggressive approach to the development and commercialization of nanotechnology. Critics of the NNI voice concerns that reflect disparate underlying beliefs. Some critics assert that the government is not doing enough to move technology from the laboratory into the marketplace. Others argue that the magnitude of the public investment may skew what should be market-based decisions in research, development, and commercialization. Still other critics say that the inherent risks of nanotechnology are not being addressed in a timely or effective manner.

Introduction

Nanotechnology has been an issue of interest to Congress for a number of years, coming into focus in 2000 with the launch of the U.S. National Nanotechnology Initiative (NNI) by President Clinton in his FY2001 budget request to Congress. From FY2001 through FY20 10, Congress appropriated approximately $12.4 billion for nanotechnology research and development (R&D).[1] These efforts have been directed at advancing understanding and control of matter at the nanoscale,[2] where the physical, chemical, and biological properties of materials differ in fundamental and useful ways from the properties of individual atoms or bulk matter.[3]

Nanotechnology: A Description

The term "nanotechnology" is often used as an all- encompassing term for nanoscale science, engineering, and technology. Nanotechnology is the understanding and control of matter at dimensions of roughly 1 to 100 nanometers, the size-scale between individual atoms and bulk materials, where unique phenomena enable novel applications. A nanometer is one-billionth of a meter, or about the width of 10 hydrogen atoms arranged side-byside in a line. Nanotechnology involves imaging, measuring, modeling, and manipulating matter at this size-scale.

At the nanoscale, the physical, chemical, and biological properties of materials can differ in fundamental and useful ways from the properties of individual atoms and molecules or bulk matter. Nanotechnology R&D is directed toward understanding and creating improved materials, devices, and systems that exploit these new properties.

Physicist Richard Feyn man's remarks at the 1959 annual meeting of the American Physical Society are often cited as the first articulation of and vision for nanotechnology. Though he did not use the term nanotechnology in this speech, he spoke of controlling matter at the nanoscale and creating atomic-level machines, positing some of the applications that doing so might enable.

Source: *The National Nanotechnology Initiative Strategic Plan, 2004,* Nanoscale Science, Engineering, and Technology Subcommittee, National Science and Technology Council, The White House, December 2004.

The development and application of nanotechnology—more fully explained below—across a wide array of products and industries holds the potential for significant economic and societal benefits. To capture these benefits, the United States will have to effectively address a variety of technical and policy challenges that stand as potential barriers to commercialization, including environmental, health, and safety (EHS) concerns and their implications for workplace, environmental, food, and drug regulations; development of standards, reference materials, and consistent nomenclature; development of new measurement methods and tools; effective technology transfer to the private sector; protection of intellectual property; availability, affordability, and patience of investment capital; ethical, legal, and societal concerns; public understanding, support, and acceptance; and development of a world-class scientific and technical nanotechnology workforce.

This chapter provides an overview of nanotechnology, the National Nanotechnology Initiative, possible reauthorization of the 21st Century Nanotechnology Research and Development Act of 2003 (P.L. 108-153), and appropriations issues.

OVERVIEW

The economic and societal promise of nanotechnology has led to involvement and investments by governments and companies around the world. In 2000, the United States became the first nation to establish a formal, national initiative to advance nanoscale science, engineering, and technology—the National Nanotechnology Initiative. Since then, Congress has appropriated approximately $12.4 billion in nanoscale science, engineering, and technology through the NNI. U.S. companies and state governments have invested billions more.

As a result of this focus and these investments, the United States has emerged as a global leader in nanotechnology. However, the competition for global leadership is intensifying as foreign investments in nanoscale science, engineering, and technology increase. Other nations have followed the U.S. lead and established their own national nanotechnology programs, each with varying degrees of investment, foci, and support for industrial applications and commercialization. Today, almost every nation that supports R&D has a national-level nanotechnology program.

Global public investments in nanotechnology in 2009 alone were estimated to be $9.8 billion, with cumulative global public investments through 2009 reaching approximately $50 billion.[4] In 2005, an assessment of the NNI by the President's Council of Advisors on Science and Technology (PCAST) reported that the United States leads all other nations in public investments in nanotechnology R&D, accounting for about a quarter of global annual public investments.[5] An April 2009 report by Cientifica, a privately-held nanotechnology business analysis and consulting firm, states that the U.S. share of global public nanotechnology funding has fallen to 19% in 2009 from 31% in 2004, calculated on a currency exchange rate basis, behind Russia (23%) and the European Union (EU) States (collectively, 27%).[6] The situation appears even less favorable when the share of investment is calculated on a purchasing power parity (PPP) basis (which takes into account the price of goods and services in each nation). Using a PPP approach, Cientifica concluded in its report that the U.S. share of global public nanotechnology investments is 16%, less than Russia (25%), and the EU States (27%), and the same as China (1 6%).[7]

Global investments in nanotechnology already have begun to yield economic benefits as products incorporating nanotechnology enter the marketplace. These products are estimated to have produced $147 billion in revenues in 2007 (including $59 billion in the United States, $47 billion in Europe, $31 billion in Asia/Pacific, and $9 billion in other nations).[8] By tapping the unique properties that emerge at the nanoscale, proponents maintain that nanotechnology holds the potential for products that could transform existing industries and create new ones, clean and protect the environment, extend and improve the quality of our lives, and strengthen national security. Most nanotechnology products currently on the market—such as faster computer processors, higher density memory devices, lighter-weight auto parts, stain-resistant clothing, antibiotic bandages, cosmetics, and clear sunscreen—are evolutionary in nature, offering incremental improvements in characteristics such as performance, aesthetics, cost, size, and weight.

Evolutionary nanotechnology products, however, represent only a small fraction of what many see as the substantial longer-term economic and societal promise of nanotechnology. One estimate projects nanotechnology product revenues will reach $3.1 trillion by 2015,[9] while another estimate projects revenues will reach $2.95 billion by 2015, of which almost half will come from semiconductors.10, [11]

Many nanotechnology advocates—including business executives, scientists, engineers, medical professionals, and venture capitalists—assert that in the longer term, nanotechnology, especially in combination with information technology, biotechnology, and the cognitive sciences, may deliver revolutionary advances, including:

- new prevention, detection, and treatment technologies that could reduce substantially death and suffering from cancer and other deadly illnesses;[12]
- new organs to replace damaged or diseased ones;[13]
- contact lenses, skin patches, and glucose-sensing tattoos that monitor diabetics' blood sugar levels and warn when too high or low;[14]
- clothing that protects against toxins and pathogens;[15]
- clean, inexpensive, renewable power through energy creation, storage, and transmission technologies;[16]
- inexpensive, portable water purification systems that provide universal access to safe water;[17]

- energy efficient, low-emission "green" manufacturing systems;[18]
- high-density memory systems capable of storing the entire Library of Congress collection on a device the size of a sugar cube;[19]
- agricultural technologies that increase crop yield and improve nutritional value, reducing global hunger and malnutrition;[20]
- self-repairing materials;[21]
- powerful, small, inexpensive sensors that can warn of minute levels of toxins and pathogens in air, soil, or water;[22] and
- decontaminated industrial sites through environmental remediation.[23]

While some applications of nanotechnology have proven market-ready, much fundamental research remains ahead, including efforts to advance understanding of nanoscale phenomena; characterize nanoscale materials; understand how to control and manipulate nanoscale particles; develop instrumentation and measurement methods; and understand how nanoscale particles interact with humans, animals, plants, and the environment. In addition, several federal agencies—such as the Departments of Defense, Energy, and Homeland Security—see the potential for nanotechnology to help address mission requirements. Historically, the federal government has played a central role in funding these types of research and development activities.

Though federal nanoscale science, engineering, and technology R&D had been underway for over a decade, the NNI was first initiated as a Presidential technology initiative in 2000.[24] The original participating agencies were the National Science Foundation (NSF), the Department of Defense (DOD), the Department of Energy (DOE), the Department of Commerce's (DOC) National Institute of Standards and Technology (NIST), the National Aeronautics and Space Administration (NASA), and the Department of Health and Human Services' National Institutes of Health (NIH). In 2009, 25 agencies participated in the NNI, including 13 that received appropriations to conduct and/or fund nanotechnology R&D. Since its first year of funding in FY200 1, the NNI's annual appropriations have grown nearly four-fold to an estimated $1.8 billion in FY20 10.[25]

In 2003, Congress provided a statutory foundation for some of the activities of the NNI through the 21st Century Nanotechnology Research and Development Act of 2003 (P.L. 108-153). The act established a National Nanotechnology Program (NNP) and provided authorizations for a subset of the NNI agencies, namely the NSF, DOE, NASA, NIST, and Environmental Protection Agency (EPA).[26] The act, however, did not address the participation of several agencies that fund nanotechnology R&D under the NNI, including DOD, NIH, and the Department of Homeland Security (DHS). Nevertheless, coordination of nanotechnology R&D activities across all NNI funding agencies continues under the National Science and Technology Council's (NSTC's) Nanoscale Science, Engineering, and Technology (NSET) subcommittee.[27] According to the NSET subcommittee's 2004 NNI Strategic Plan, "For continuity and to capture this broader participation, the coordinated federal activities as a whole will continue to be referred to as the National Nanotechnology Initiative." Accordingly, the functions and activities established under the act are incorporated into the Executive Branch's implementation of the NNI.

The thrust of the NNI has primarily been the development of fundamental scientific knowledge through basic research. Investments at mission agencies, such as DOD, have supported nanotechnology applications development for which they are a primary customer.

Other investments have supported infrastructural technologies. For example, NIST has contributed to developing tools and standards that enable measurement and control of matter at the nanoscale, thereby supporting the conduct of R&D and the ability to manufacture nanoscale materials and products. As understanding of nanotechnology has matured, the NNI has worked with a variety of industry organizations to facilitate the movement of research results from the laboratory bench to the marketplace in fields as disparate as semiconductors, chemicals, energy, concrete, and forest products.

The NNI agencies also have begun to address research needs and regulatory issues related to environmental, health, and safety issues, as well as issues such as public understanding and workforce education and training. The NNI agencies actively engage in a variety of international fora, such as the Organization for Economic Cooperation and Development (OECD) and the International Standards Organization (ISO), to cooperatively address nanotechnology issues related to EHS, metrology[28] and standards, nomenclature, and nanoscale materials characterization.

Maintaining U.S. leadership poses a variety of technical, economic, and policy challenges, including:

- safeguarding the environment and ensuring human health and safety;
- creating the standards, reference materials, nomenclature, methods, and tools for metrology to enable the manufacturing of nanoscale materials and products;
- developing a world-class scientific and technical nanotechnology workforce;
- translating research results into products, including effective technology transfer to the private sector;
- understanding public perceptions and attitudes and fostering public understanding;
- addressing ethical, legal and societal implications;
- protecting intellectual property;
- securing investment capital for early-stage research, development, and commercialization; and
- fostering and facilitating international cooperation and coordination.

Proponents of the NNI assert that nanotechnology is one of the most important emerging and enabling technologies[29] and that U.S. competitiveness, technological leadership, national security, and societal interests require an aggressive approach to the development and commercialization of nanotechnology. Critics of the NNI hold a variety of competing views, asserting that government is not doing enough, is doing too much, or is moving too quickly.

Some in industry have criticized the NNI for being overly focused on basic research and not being aggressive enough in moving NNI-funded R&D out of government and university laboratories and into industry. Others in industry have criticized the federal government for not providing mechanisms to help advance nanotechnology R&D to the point where it becomes economically viable for venture capitalists, corporations, and other investors to create products and bring them to market. Some refer to this gap as the "valley of death."[30] Still others in industry have criticized the NNI for not adequately supporting the development of metrology, standards, equipment, and processes necessary to manufacture nanotechnology materials, products, and systems at a commercial scale.

Conversely, supporters of industry-driven market investments contend that extensive government support for nanotechnology may supplant the judgment of the marketplace by picking "winners and losers" in technological development. For example, the size and directions of the NNI investments may encourage industry to follow the government's lead rather than independently selecting R&D directions itself or, alternatively, may result in the promotion of a less effective technology path over a more effective one. These supporters also assert that federal government funding of scientific research is often wasteful, driven by political considerations and not scientific merit.[31]

Some non-governmental organizations (NGO) are critical of nanotechnology for its potential adverse impacts on human health and safety and on the environment. They assert that the government is pushing ahead too quickly in developing nanotechnology and encouraging its commercialization and use without adequately investing in research focused on understanding and mitigating negative EHS implications.[32] They argue that the very characteristics that make nanotechnology promising also present significant potential risks to human health and safety and the environment. Some of these critics argue for application of the "precautionary principle," which holds that regulatory action may be required to control potentially hazardous substances even before a causal link has been established by scientific evidence.[33] At least one NGO has called for a moratorium on nanotechnology R&D and new commercial products incorporating synthetic nanoparticles.[34]

NATIONAL NANOTECHNOLOGY INITIATIVE

The National Nanotechnology Initiative is an interagency program that coordinates federal nanoscale science, engineering, and technology R&D activities and related efforts among participating agencies.

Vision and Goals

The National Science and Technology Council (NSTC) has stated the following vision for the NNI:

> A future in which the ability to understand and control matter on the nanoscale leads to a revolution in technology and industry. The NNI will expedite the discovery, development, and deployment of nanotechnology in order to achieve responsible and sustainable economic benefits, to enhance the quality of life, and to promote national security.[35]

To achieve its vision, the NNI has established four goals: maintain a world-class R&D program aimed at realizing the full potential of nanotechnology; facilitate transfer of new technologies into products that provide economic growth, jobs, and other public benefits; develop educational resources, a skilled workforce, and the supporting infrastructure and tools to advance nanotechnology; and support responsible development of nanotechnology.[36]

History

Attempts to coordinate federal nanoscale R&D began in November 1996, as staff members from several agencies met regularly to discuss their plans and programs in nanoscale science and technology. This group continued informally until September 1998, when it was designated as the Interagency Working Group on Nanotechnology (IWGN) under the NSTC. In August 1999, IWGN completed its first draft of a plan for an initiative in nanoscale science and technology, which was subsequently approved by the President's Council of Advisors on Science and Technology (PCAST) and the White House Office of Science and Technology Policy (OSTP).[37]

In his 2001 budget submission to Congress, then-President Clinton raised nanotechnology-related research to the level of a federal initiative, officially referring to it as the National Nanotechnology Initiative. [38]

Legislative Approach

Congress has played a central role in the National Nanotechnology Initiative, providing appropriations for the conduct of nanoscale science, engineering, and technology research; establishing programs; and creating a legislative foundation for the activities of the NNI.

Congressional funding for the NNI is provided through appropriations to each of the NNI-participating agencies. The NNI has no centralized funding. The overall NNI budget is calculated by aggregating the nanotechnology budgets for each of the federal agencies that conduct or provide funding for nanoscale science, engineering, and technology research.

In FY200 1, the first year of NNI funding, Congress provided $464 million to eight agencies for nanoscale science, engineering, and technology research.[39] The NNI has continued to receive support from both Congress and the White House. Both the number of agencies participating in the NNI and the size of the federal investment have grown. Currently 25 agencies participate in the NNI, 13 of which have received appropriated funds for nanotechnology R&D in FY2009.[40] Total NNI funding in FY2009 is approximately $1.8 billion, more than three times the level of funding provided in FY200 1. The original six agencies identified at the launch of the NNI[41] still account for the vast majority of NNI funding (97.3% in FY2009; detailed agency funding levels for FY20 10 are not yet available).

21st Century Nanotechnology Research and Development Act of 2003

Congress codified and further defined some of the NNI's activities in the 21st Century Nanotechnology Research and Development Act of 2003 which was passed by Congress in November 2003, and signed into law (P.L. 108-153) by President Bush on December 3, 2003.[42] The legislation received strong bipartisan support in both the House of Representatives, which passed the bill on a recorded vote of 405-19, and in the Senate, which passed the bill by unanimous consent.

Though this act is often referred to as the enabling legislation for the National Nanotechnology Initiative, the act actually establishes a National Nanotechnology Program (NNP). The act provides authorizations for five NNI agencies—the National Science Foundation, Department of Energy, NASA, National Institute of Standards and Technology, and Environmental Protection Agency—but not for the Department of Defense, National Institutes of Health, Department of Homeland Security,[43] or other NNI research agencies that collectively accounted for 46% of NNI funding in FY2003.

The act created the NNP for the purposes of establishing the goals, priorities, and metrics for evaluation of federal nanotechnology research, development, and other activities; investing in federal R&D programs in nanotechnology and related sciences to achieve those goals; and providing for interagency coordination of federal nanotechnology research, development, and other activities undertaken pursuant to the NNP.

Key provisions of the act include:

- authorizing appropriations for the nanotechnology-related activities of the National Science Foundation, Department of Energy, NASA, National Institute of Standards and Technology, and Environmental Protection Agency for fiscal years 2005 through 2008, totaling $3.679 billion for the four year period;
- establishing a National Nanotechnology Coordination Office, with a director and full time staff to provide administrative support to the NSTC;
- establishing a National Nanotechnology Advisory Panel (NNAP) to advise the President and the NSTC on matters relating to the NNP.
- establishing a triennial review of the NNP by the National Research Council of the National Academies of Sciences;
- directing the NSTC to oversee the planning, management, and coordination of the program, including the development of a triennial strategic plan;
- directing the Department of Commerce's National Institute of Standards and Technology to establish a program to conduct basic research on issues related to the development and manufacture of nanotechnology, and to use the Manufacturing Extension Partnership program to ensure results reach small- and medium-sized manufacturing companies;
- directing the Secretary of Commerce to use the National Technical Information Service to establish a clearinghouse of information related to commercialization of nanotechnology research;
- directing the Secretary of Energy to establish a program to support consortia to conduct interdisciplinary nanotechnology R&D designed to integrate newly developed nanotechnology and microfluidic tools with systems biology and molecular imaging;
- directing the Secretary of Energy to carry out projects to develop, plan, construct, acquire, operate, or support special equipment, instrumentation, or facilities for investigators conducting nanotechnology R&D; and
- directing the establishment of two centers, on a merit-reviewed and competitive basis: (1) the American Nanotechnology Preparedness Center, to conduct, coordinate, collect, and disseminate studies on the societal, ethical, environmental, educational, legal, and workforce implications of nanotechnology; and to identify anticipated issues related to the responsible research, development, and application of nanotechnology, as well as provide recommendations for preventing or addressing such issues, and (2) the Center for Nanomaterials Manufacturing, to encourage, conduct, coordinate, commission, collect, and disseminate research on new manufacturing technologies for materials, devices, and systems with new combinations of characteristics, such as, but not limited to, strength,

toughness, density, conductivity, flame resistance, and membrane separation characteristics; and to develop mechanisms to transfer such manufacturing technologies to U.S. industries.

While the act establishes a National Nanotechnology Program, the Executive Branch continues its broader effort under the NNI framework and name. According to the NNI's 2004 Strategic Plan:

> Many of the activities outlined in the Act were already in progress as part of the NNI. Moreover, the ongoing management of the initiative involves considerable input from Federal agencies that are not named specifically in the Act.... For continuity, and to capture this broader participation, the coordinated Federal activities as a whole will continue to be referred to as the National Nanotechnology Initiative.[44]

Reauthorization Efforts

The 21st Century Nanotechnology Research and Development Act provided a legislative foundation for some of the activities of the NNI, authorized agency funding levels through FY2008, and sought to address challenges associated with the development and commercialization of nanotechnology. While many provisions of this act have no sunset provision, FY2008 was the last year of agency authorizations included in the act. Legislation to amend and reauthorize the act was introduced in the House (H.R. 5940, 110th Congress) and the Senate (S. 3274, 110th Congress) in the 110th Congress. Both bills were titled the National Nanotechnology Initiative Amendments Act of 2008. The House passed H.R. 5940 by a vote of 407-6; the Senate did not act on S. 3274.

In January 2009, H.R. 554, the National Nanotechnology Initiative Amendments Act of 2009, was introduced in the 111th Congress. The act contains essentially the same provisions as H.R. 5940 (1 10th Congress). (For additional information on H.R. 554, see bill highlights on page 38.) In February 2009, the House passed the bill by voice vote under a suspension of the rules. In July 2009, S. 1482, the National Nanotechnology Initiative Amendments Act of 2009, was introduced in the Senate. The bill was referred to the Senate Committee on Commerce, Science, and Transportation. No further action has been taken. H.R. 820, the Nanotechnology Advancement and New Opportunities Act, also would amend P.L. 108-153. The provisions of H.R. 820 cover a variety of jurisdictions, thus the bill has been assigned to multiple House committees. On May 7, 2010, the House Committee on Science and Technology reported the America COMPETES Reauthorization Act of 2010 (H.R. 5116) which includes, as Title I, Subtitle A, the "National Nanotechnology Initiative Amendments Act of 2010." Provisions of this subtitle are nearly identical to the provisions of H.R. 554. (For additional information on these bills, see "Selected Nanotechnology Legislation in the 111th Congress.")

Congress may use these bills to further address these issues and to establish authorization levels for agency nanotechnology R&D. Alternatively, Congress may choose to address some or all of these issues in separate legislation. Several bills have been introduced in the first session of the 111th Congress to address specific nanotechnology issues.

Structure

Nanoscale Science, Engineering, and Technology Subcommittee

The NNI is coordinated within the White House through the NSTC, the Cabinet-level council by which the President coordinates science, space, and technology policies across the federal government. Operationally, NNI coordination is accomplished through the Nanoscale Science, Engineering, and Technology (NSET) subcommittee of the NSTC's Committee on Technology (CT). The NSET subcommittee also has an informal reporting relationship to the NSTC's Committee on Science (CS). The NSET subcommittee is led by an agency co-chair, currently from the Department of Energy (DOE), and an OSTP co-chair. The NSET subcommittee is comprised of representatives from 25 federal entities (including 15 that fund nanotechnology R&D), OSTP and the Office of Management and Budget.[45]

The NSET subcommittee has established several working groups, each taking on efforts in key subject areas.[46] Among them:

National Environmental and Health Implications (NEHI)

The National Environmental and Health Implications (NEHI) working group was chartered to provide for exchange of information among agencies that support research and those responsible for regulations and guidelines related to nanotechnology products; to facilitate identification, prioritization, and implementation of research and other activities required for the responsible research, development, utilization, and oversight of nanotechnology; and to promote communication of information related to research on environmental and health implications of nanotechnology to other government agencies and non-government parties. To this end, the NEHI working group seeks to identify and prioritize environmental, health, and safety research needs related to nanotechnology. Twenty NNI agencies participate in the NEHI working group, and 13 agencies fund safety-related nanotechnology research and/or have regulatory authorities to guide the safe use of nanomaterials.[47]

Nanomanufacturing, Industry Liaison, and Innovation(NILI)

The Nanomanufacturing, Industry Liaison, and Innovation (NILI) working group was chartered to enhance collaboration and information sharing between U.S. industry and government on nanotechnology-related activities. It also facilitates federal, regional, state, and local nanotechnology R&D and commercialization activities. In addition, the NILI working group is to create innovative methods for transferring federally funded technology to industry. The NILI working group has facilitated collaborations between the NNI and the semiconductor/electronics industry, chemical industry, forest products industry, and the Industrial Research Institute.[48]

Global Issues in Nanotechno logy (GIN)

The Global Issues in Nanotechnology (GIN) working group was chartered to monitor foreign nanotechnology programs and development; broaden international collaboration on nanotechnology R&D, including safeguarding the environment and human health; and promote U.S. commercial and trade interests in nanotechnology. The NEHI working group works with the GIN working group to coordinate the U.S. position and participation in international activities related to environmental, health, and safety implications of nanotechnology. The GIN working group facilitates international collaboration on pre-competitive and non-competitive aspects of nanotechnology, and international engagement on trade, commercialization and regulatory issues.

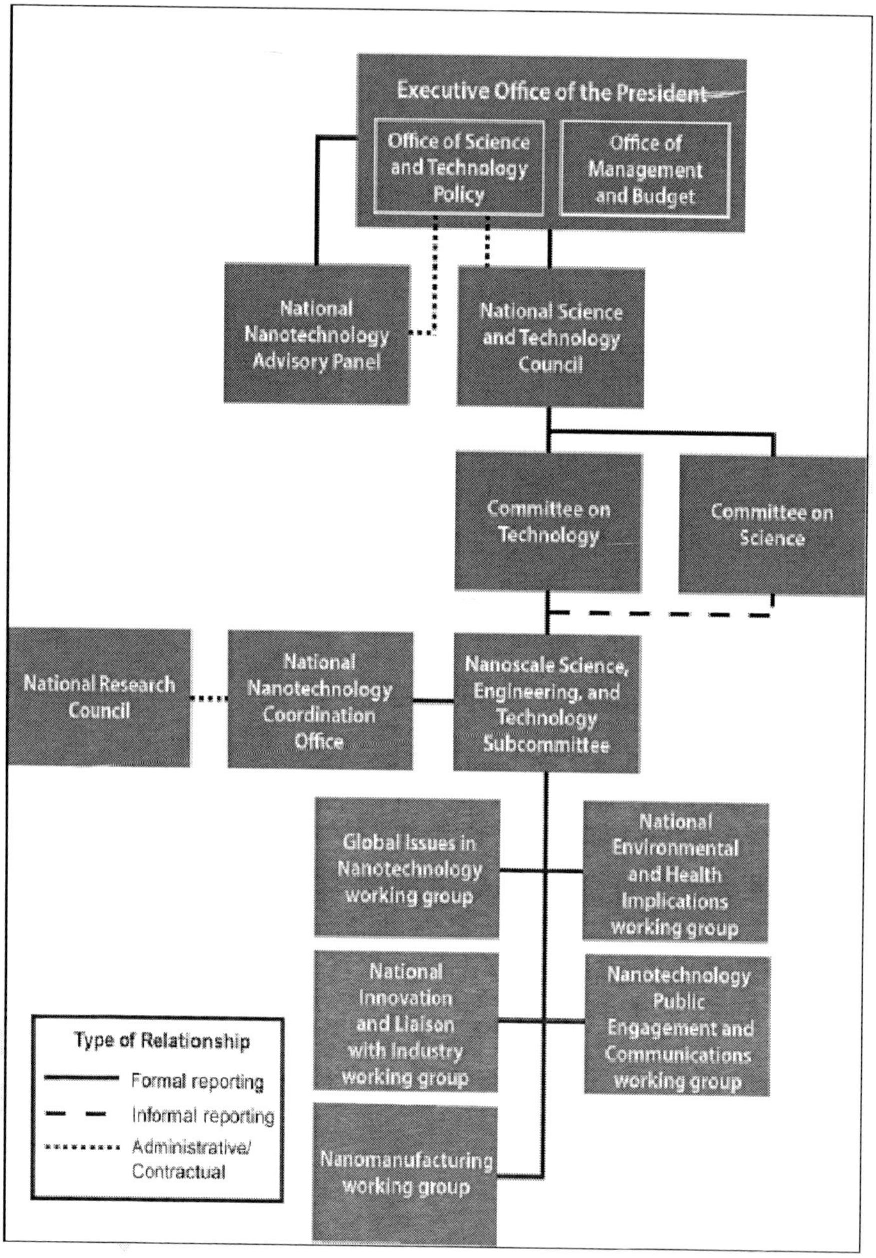

Source: *The National Nanotechnology Strategic Plan*, Nanoscale, Science, Engineering, and Technology Subcommittee, National Science and Technology Council, The White House, December 2007. (Nanomanufacturing working group added to chart in cited source.)

Figure 1. Organizations With a Role in the National Nanotechnology Initiative and Their Relationships

Nanotechnology Public Engagement and Communications (NPEC)

The Nanotechnology Public Engagement and Communications (NPEC) working group was established to develop approaches by which the NNI can communicate more effectively with the public.

National Nanotechnology Coordination Office

The National Nanotechnology Coordination Office (NNCO) provides administrative and technical support to the NSET subcommittee. Initially established in 2001 through a memorandum of understanding among the NNI participating agencies,[49] the NNCO was authorized by the 21st Century Nanotechnology Research and Development Act of 2003 (P.L. 108-153). The NNCO was charged under the act with providing technical and administrative support to the NSTC and NNAP; serving as the point of contact for information on Federal nanotechnology activities for the exchange of technical and programmatic information among stakeholders; conducting public outreach; and promoting access to and early application of NNP technologies, innovation, and expertise.

The act authorizes the work of the NNCO to be funded by contributions from NSET subcommittee member agencies. According to the NNCO, funding is provided through a memorandum of understanding signed by eight NNI agencies.[50] In principle, each agency contributes to the NNCO budget in proportion to its share of the President's total nanotechnology budget request for the signatory agencies. However, two of the signatories, EPA and DOT, had sufficiently small enough nanotechnology budgets in the early years of the NNI that they were not expected to contribute. EPA now contributes to funding the NNCO. Total NNCO funding from the agencies in FY20 10 is $2.4 million. In addition, the NNCO has carried over FY2009 funds intended to support a National Academies' study, bringing the total NNCO budget for FY20 10 to just under $3 million.

Funding

The NNI supports fundamental and applied research on nanotechnology by funding research, creating multidisciplinary centers of excellence, and developing key research infrastructure. It also supports activities aimed at addressing the societal implications of nanotechnology, including ethical, legal, human and environmental health, and workforce issues.

This section provides information on NNI funding from two perspectives: organizationally by agency and functionally by program component area.

Agency Funding

The NNI budget is an aggregation of the nanotechnology components of the individual budgets of NNI-participating agencies. The NNI budget is not a single, centralized source of funds that is allocated to individual agencies. In fact, agency nanotechnology budgets are developed internally as part of each agency's overall budget development process. These budgets are subjected to review, revision, and approval by the Office of Management and Budget and become part of the President's annual budget submission to Congress. The NNI budget is then calculated by aggregating the nanotechnology components of the appropriations provided by Congress to each federal agency.

For FY20 10, regular appropriations for NNI agencies totaled an estimated $1.78 1 billion,[51] a $79.5 million (4.7%) increase over the FY2009 regular appropriation and nearly four times the $464 million federal investment in nanotechnology research in FY200 1. This growth in nanotechnology R&D investments reflects expectations in Congress and in the executive branch that the NNI will expand fundamental knowledge and make important contributions to national priorities. The chronology of NNI funding is detailed in **Table 1**.

President Obama has requested $1 .762 billion in funding for the NNI in FY20 11, a decrease of $19.5 million (1.1%). The FY20 11 NNI budget request would support a broad range of programs among 13 agencies. Agencies with the largest budgets are:

- NSF, which supports fundamental nanotechnology research across science and engineering disciplines;
- DOD, whose investments in nanotechnology are aimed at addressing the department's national security mission;
- DOE, which supports nanotechnology research providing a basis for new and improved energy efficiency, production, storage, and transmission technologies;
- NIH, which emphasizes nanotechnology-based biomedical advances occurring at the intersection of biology and the physical sciences; and
- NIST, which focuses on research in instrumentation, measurement, standards, characterization, and nanomanufacturing.

Other agencies investing in mission-related nanotechnology R&D are NASA, EPA, the Cooperative State Research, Education, and Extension Service (CSREES) and Forest Service at the Department of Agriculture (USDA), National Institute of Occupational Safety and Health (NIOSH), DHS, Department of Justice (DOJ), and Department of Transportation's (DOT's) Federal Highway Administration (FHWA).

Program Component Area Funding

The 21st Century Nanotechnology R&D Act of 2003 called for the NSET subcommittee to develop categories of investment called Program Component Areas (PCA) to provide a means by which Congress and the executive branch can be informed of and direct the relative investments in these areas. The PCAs are categories of investments that cut across the needs and interests of individual agencies and contribute to the achievement of one or more of the NNI's goals. The 2004 NNI strategic plan identified seven PCAs. The 2007 NNI strategic plan splits the seventh PCA, Societal Dimensions, into two PCAs: Environment, Health, and Safety; and Education and Societal Dimensions. A description of the seven initial PCAs and their current funding are provided below,[52] as well as a description of the two derivative PCAs.[53] The chronology of NNI funding by PCA is detailed in **Table 2**.

In the following analysis of funding for each of the PCAs, comparisons involving FY2009 funding (i.e., between FY20 10 and FY2009) exclude ARRA funding. In addition, comparisons between the President's FY20 11 budget request and estimated FY20 10 appropriations may be affected by the President's exclusion of congressionally directed funding in his budget.

Table 1. NNI Funding, by Agency: FY200 1 -FY2009 and FY20 10 Request (in Millions of Current Dollars)

Agency	FY 2001 Actual	FY 2002 Actual	FY 2003 Actual	FY 2004 Actual	FY 2005 Actual	FY 2006 Actual	FY 2007 Actual	FY 2008 Actual	FY 2009 Actual	ARRA (P.L. 111-5)[a]	FY2010 Estimate	FY2011 Request
National Science Foundation	150	204	221	256	335	360	389	409	409	101	418	401
Department of Defense[b]	125	224	220	291	352	424	450	460	459		436	349
Department of Energy	88	89	134	202	208	231	236	245	333	293	373	424
National Institutes of Health (HHS)	40	59	78	106	165	192	215	305	343	73	361	382
Nat'l Inst. of Standards and Tech. (DOC)	33	77	64	77	79	78	88	86	93	43	114	108
Environmental Protection Agency	5	6	5	5	7	5	8	12	12		18	20
NASA	22	35	36	47	45	50	20	17	14		14	16
Department of Justice	1	1	1	2	2	<1	2	0	1		0	
Department of Homeland Security		2	1	1	1	2	2	3	9		12	12
Nat'l Inst. of Food and Agriculture (USDA)[c]			1	2	3	4	4	6	10		10	9
Nat'l Inst. for Occupational Safety (HHS)					3	4	7	7	7		10	17
Forest Service (USDA)						2	3	5	5		5	5

Table 1. (Continued)

Agency	FY 2001 Actual	FY 2002 Actual	FY 2003 Actual	FY 2004 Actual	FY 2005 Actual	FY 2006 Actual	FY 2007 Actual	FY 2008 Actual	FY 2009 Actual	ARRA (P.L. 111-5)[a]	FY2010 Estimate	FY2011 Request
Federal Highway Administration (DOT)						1	1	1	1		3	2
Food and Drug Administration (HHS)									7		7	15
Consumer Product Safety Commission									<1		<1	2
TOTAL[d]	464	697	760	989	1,200	1,351	1,425	1,554	1,702	511	1,781	1,762

Sources: NNI website, http://www.nano.gov.

a. Funding figures for nanotechnology-related R&D under the ARRA are preliminary estimates.

b. According to NSTC, the DOD budgets shown above for FY2006, FY2007, FY2008, and FY2009 include congressionally directed funding of approximately $76 million in FY2006, $63 million in FY2007, and $117 million in FY2009; the NSTC states that the DOD budget for FY20 10 also includes congressionally directed funding but does not provide an amount. According to NSTC, the 2008 DOD estimate "includes many earmarks that are outside the NNI plan."

c. Formerly, the USDA Cooperative State Research, Education, and Extension Service (CSREES).

d. Totals may not add due to rounding of agency budget figures.

Table 2. NNI Funding, by Program Component Area, FY2006-FY2009 (in Millions of Current Dollars)

PCA	FY2006 Actual	FY2007 Actual	FY2008 Actual	FY2009 Actual	ARRA Preliminary	FY2010 Estimate	FY2011 Request
Fundamental Phenomena and Processes	455.9	480.6	478.5	479.2	130.6	467.9	484.4
Nanomaterials	265.1	258.3	285.1	331.9	178.3	373.5	342.3
Nanoscale Devices and Systems	319.6	344.7	372.7	435.2	68.0	429.9	402.0
Instrumentation Research, Metrology, and Standards	51.0	52.5	69.0	90.8	12.4	84.3	76.9
Nanomanufacturing	33.8	48.1	47.1	75.6	28.5	96.7	101.4
Major Research Facilities and Instrumentation Acquisition	152.4	152.4	196.4	177.6	72.5	197.3	203.0
Societal Dimensions	73.5						
• Environment, Health, and Safety		48.3	67.9	74.5	12.0	91.6	116.9
• Education & Societal Dimensions		39.2	37.7	36.8	9	39.9	34.8
TOTAL[a]	**1,351.2**	**1,424.1**	**1,554.4**	**1,701.5**	**511.3**	**1,781.1**	**1,761.6**

Source: *The National Nanotechnology Initiative: Research and Development Leading to a Revolution in Technology and Industry, Supplement to the President's FY2008 Budget*, Nanoscale Science, Engineering, and Technology Subcommittee, *National Science and Technology Council, The White House, July 2007. National Nanotechnology Initiative: FY2009 Budget & Highlights, National Science and Technology Council*, The White House, February 2008. T*he National Nanotechnology Initiative: Research and Development Leading to a Revolution in Technology and Industry, Supplement to the President's FY2010 Budget, Nanoscale Science, Engineering, and Technology Subcommittee*, National Science and Technology Council, The White House, May 2009.

a. Details may not add to totals due to rounding.

Fundamental Phenomena and Processes

Fundamental Phenomena and Processes includes investments in the discovery and development of fundamental knowledge pertaining to the new phenomena in the physical, biological, and engineering sciences that occur at the nanoscale, as well as in understanding and articulation of scientific and engineering principles related to nanoscale structures, processes, and mechanisms.

FY20 10 funding for Fundamental Phenomena and Processes fell to $467.9 million, down $11.3 million (2.4%) from the FY2009 level due primarily to a decrease in DOD funding in this PCA (down $24.0 million, 14.7%). President Obama's FY20 11 budget proposes $484.4 million in funding, up $16.5 million (3.5%) over the FY20 10 level due primarily to increases in DOE ($14.2 million, 13.8%) and DOD ($412.7 million, 9.1%) funding. The increase in DOE and DOD funding would somewhat offset by a decrease in NSF funding of $12.5 million (8.2%).[54]

Nanomaterials

Nanomaterials includes research investments to discover novel nanoscale and nanostructured materials. This PCA also attempts to understand the properties of nanomaterials, and supports R&D to enable the design and synthesis, in a controlled manner, of nanoscale materials with targeted properties.

FY20 10 funding for Nanomaterials rose to $373.5 million, up $41.6 million (12.5%) from the FY2009 level, led by an increase in DOE funding in this PCA (up $21.7 million, 23.4%). President Obama's FY20 11 budget proposes $342.3 million for this PCA, a decrease of $31.2 million (8.4%) from the FY20 10 level, resulting primarily from reduced DOD funding (down $36.0 million, 47.8%).[55]

Nanoscale Devices and Systems

Nanoscale Devices and Systems include R&D investments that apply nanoscale science and engineering principles to create novel devices and systems or to improve existing ones. It also includes the use of nanoscale or nanostructured materials to achieve improved performance or new functionality. To meet this definition, the enabling science and technology must be at the nanoscale, but the systems and devices are not restricted to that size.

Funding for Nanoscale Devices and Systems fell to $429.9 million in FY2010, down $5.3 million (1.2%) from the FY2009 level. President Obama's FY20 11 budget proposes $402.0 million in funding for this PCA, a decrease of $27.9 million (6.5%) from the FY20 10 level, largely due to reductions in DOD funding (down $49.0 million, 33.1%). The decrease in DOD funding for this PCA would somewhat offset by increases in DOE ($13.2 million, 76.7%) and NIH funding ($13.1 million, 7.2%).[56]

Instrumentation Research, Metrology, and Standards

The Instrumentation Research, Metrology, and Standards PCA includes R&D investments for development of tools needed to advance nanotechnology research and commercialization. Instrumentation for characterization, measurement, synthesis, and design of nanotechnology materials, structures, devices, and systems is funded through this PCA.

R&D and other activities related to development of standards, including standards for nomenclature, materials, characterization, testing, and manufacture are also in this PCA.

FY20 10 funding for Instrumentation Research, Metrology, and Standards fell to $84.3 million, down $6.5 million (7.2%) from the FY2009 level. President Obama's FY20 11 budget proposes $76.9 million in funding for this PCA, a decrease of $7.4 million (8.8%) from the FY20 10 level, due to decreases in DOD, DOE, and NSF funding.[57]

Nanomanufacturing

Nanomanufacturing R&D supports the development of scalable, reliable, cost-effective manufacturing of nanoscale materials, structures, devices, and systems. It also includes R&D and integration of ultra-miniaturized top-down processes and complex bottom-up processes.[58]

FY20 10 funding for Nanomanufacturing rose to $96.7 million, up $21.1 million (27.9%) over the FY2009 level due primarily to increases in NIST ($17.8 million, 189.4%) and DOD ($8.2 million, 28.3%) funding. President Obama's FY20 11 budget proposes $101.4 million for this PCA, an increase of $4.7 million (4.9%) above the FY2010 level as increases in DOE ($13.9 million, 198.6%) and NSF ($9.8 million, 43.8%) would more than offset decreases in DOD ($12.1 million, 32.5%) and NIST ($7.0 million, 25.7%) funding.[59]

Major Research Facilities and Instrumentation Acquisition

This PCA includes investments in the establishment and ongoing operations of user facilities and networks, the acquisition of major instrumentation, and other activities related to infrastructure for the conduct of nanoscale science, engineering, and technology R&D.

FY20 10 funding for Major Research Facilities and Instrumentation Acquisition rose to $197.3 million, an increase of $19.7 million (11.1%) over the FY2009 level. President Obama's FY20 11 budget proposes $203.0 million for this PCA, an increase of $5.7 million (2.9%) above the FY2009 level, led by increases in DOE and DOD funding.[60]

Societal Dimensions

The Societal Dimensions PCA includes investments in research and other activities that address the broad implications of nanotechnology to society. This includes assessing benefits and risks through research directed at environmental, health, and safety impacts of nanotechnology development; risk assessment of such impacts; education-related activities, such as development of materials for schools, undergraduate programs, technical training, and public outreach; and research directed at identifying and quantifying the broad implications of nanotechnology for society, including social, economic, workforce, educational, ethical, and legal implications.

Under the 2007 NNI Strategic Plan, the Societal Dimensions PCA was divided into two separate PCAs: Environment, Health, and Safety, and Education and Societal Dimensions. Future PCA reporting will use the new eight PCA taxonomy. NSTC has retroactively reported FY2007 Societal Dimensions PCA spending in the new PCAs.[61] The NSET subcommittee characterizes the new PCAs as follows:[62]

Environment, Health, and Safety

This PCA addresses research primarily directed at understanding the environmental, health, and safety impacts of nanotechnology development and corresponding risk assessment, risk management, and methods for risk mitigation.

FY20 10 funding for Environment, Health, and Safety rose to $91.6 million, up $17.1 million (23.0%) above the FY2009 level, due to increases in EPA, NIH, NSF, and NIOSH funding. President Obama's FY20 11 budget proposes $116.9 million in funding for this PCA, an increase of $25.3 million (27.6%) above the FY20 10 level, led by increases at FDA and NIOSH.[63]

Education and Societal Dimensions

This PCA addresses education-related activities such as development of materials for schools, undergraduate programs, technical training, and public communication, including outreach and engagement. Such activities include research directed at identifying and quantifying the broad implications of nanotechnology society, including social, economic, workforce, educational, ethical, and legal implications.

FY20 10 funding for Education and Societal Dimensions rose slightly to $39.9 million, up $3.1 million (8.4%) from the FY2009 level due primarily to an increase in NSF funding ($3.0 million, 9.6%); NSF accounts for 86.0% of funding in this PCA in FY2009. President Obama's FY201 1 budget proposes $34.8 million in funding for this PCA, a decrease of $5.1 million (12.8%) from the FY2010 level, accounted for entirely by a decrease in NSF funding.[64]

Centers, Networks, and User Facilities

A key facet of the National Nanotechnology Initiative has been the development of an extensive infrastructure of interdisciplinary research and education centers, networks, and user facilities. The centers and user facilities are located at universities and federal laboratories across the country.

Centers and networks provide opportunities and support for multidisciplinary research among investigators from a variety of disciplines and research sectors, including academia, industry, and government laboratories. Such multidisciplinary research not only can lead to advances in knowledge, but also may foster relationships that further the development of basic research results into devices and other applications.

Many agencies support such centers. Examples of federal and federally supported centers include:

- The National Science Foundation has established university-based centers focused exclusively on nanotechnology, including 15 Nanoscale Science and Engineering Centers (NSEC), one Engineering Research Center, one Science and Technology Center, four Materials Research Science and Engineering Centers, two Nanoscale Science and Engineering Education Centers, and five Nanoscale Science and Engineering Networks.[65]
- The NIH has established more than 20 centers, including eight university-based Nanomedicine Development Centers; a Nanotechnology Characterization Laboratory,

established by the National Cancer Institute (NCI), in partnership with NIST and the Food and Drug Administration; eight university-based Centers of Cancer Nanotechnology Excellence, established under the NCI's Alliance for Nanotechnology in Cancer initiative; and four university-based centers, established by the National Heart, Lung, and Blood Institute under its Program of Excellence in Nanotechnology.

- The Department of Defense supports two university-based nanotechnology research centers, as well as the Institute for Nanoscience at the Naval Research Laboratory.
- NASA has established three centers under its University Research, Engineering, and Technology Institute program.
- The Department of Energy has established five Nanoscale Science Research Centers (NSRCs) co-located with its national labs.
- NIST has established a Center for Nanoscale Science and Technology (CNST).
- NIOSH has established a Nanotechnology Research Center to conduct research into the application of nanoparticles and nanomaterials in occupational safety and health and the implications of nanoparticles and nanomaterials for work-related injury and illness.

Many of the centers are designated as user facilities and are available to researchers not located at the center. User facilities are designed to allow outside researchers to take advantage of facilities, equipment, tools, and expertise. These shared resources provide researchers the opportunity to conduct research, characterize materials, and test products using equipment and facilities that their individual companies, universities, or organizations could not afford to acquire, support, or maintain. Conditions for user access vary by facility and agency. In general, users are not charged for pre-competitive, non-proprietary work leading to publication, and are charged on a cost- recovery basis for proprietary work. In some cases, the user facilities are located at federal government laboratories (e.g. the Department of Energy's five NSRCs, and the NIST CNST); other user facilities are located at universities and supported with federal funds (e.g. NSF's 13 university-based centers in the National Nanotechnology Infrastructure Network (NNIN)).

As mentioned earlier, the 21st Century Nanotechnology R&D Act of 2003 directed the establishment of two centers, the American Nanotechnology Preparedness Center and the Center for Nanomaterials Manufacturing. According to the NSET subcommittee, the requirement to establish the American Nanotechnology Preparedness Center was met by NSF 's establishment of the Network for Nanotechnology in Society, comprised of centers at the University of California, Santa Barbara (with the participation of Harvard University and the University of South Carolina) and the University of Arizona.[66] These centers were funded under NSF's Nanoscale Science and Engineering Center (NSEC) program and did not include participation by any other NSET subcommittee agency. [67] The NSET subcommittee states that the requirement for establishing the Center for Nanomaterials Manufacturing was met by NSF 's establishment of a National Nanomanufacturing Network (NNN) comprised of four NSECs. The Center for Integrated Hierarchical Manufacturing at the University of Massachusetts Amherst is the main node of the NNN.[68] The NNN NSECs were established by NSF in collaboration with DOD and NIST, but exclusively with NSF funds.[69]

SELECTED NNI REPORTS AND ASSESSMENTS

This section presents summaries of recent reports from the NSTC's Nanoscale Science, Engineering, and Technology Subcommittee and assessments conducted by the National Research Council and the President's Council of Advisors on Science and Technology.

Selected NNI Reports

The NNI's coordinating body, the NSTC's Nanoscale Science, Engineering, and Technology Subcommittee, produces a variety of reports that serve to inform Congress and other key stakeholders on the initiatives' current activities, investments, and priorities.

The National Nanotechnology Initiative: Research and Development Leading to a Revolution in Technology and Industry, Supplement to the President's FY2011 Budget[70]

Each year the NSET subcommittee publishes a supplement to the President's annual budget request. The FY2011 NNI budget supplement provides a more detailed look at NNI funding in the President's budget request, including a break-out of the prior, current, and requested year budgets for each PCA. The report describes proposed changes in agency R&D budgets, as well as in the balance of investments by PCA. Of particular note:

- President Obama has proposed an overall NNI budget for FY2011 of $1.762 billion, a $19.5 million (1.1%) decrease below FY2010 funding.
- Estimated funding for EHS R&D in FY2010 rose to $91.6 million, a 23.0% increase over FY2009. President Obama has requested $116.9 million for EHS R&D in FY2011, an increase of $25.3 million (27.6%) over FY2010.
- President Obama's FY2011 budget request reflected a decline in spending in several program component areas compared to FY2010 estimated funding levels, including nanomaterials, down $12.0 million (3.9%); nanoscale devices and systems, down $31.2 million (8.4%); nanoscale devices and systems, down $27.9 million (6.5%), and instrumentation research, metrology and standards, down $7.4 million (8.8%). In addition to EHS funding, other program component areas that would be increased include fundamental phenomena and processes, up $16.5 million (3.5%), major research facilities and instrumentation acquisition, up $5.7 million (2.9%), and nanomanufacturing, up $4.7 million (4.9%%).

The National Nanotechnology Strategic Plan (2007)[71]

The 21st Century Nanotechnology R&D Act of 2003 (P.L. 108-153) requires the NSTC to develop an NNI strategic plan every three years. This plan is to guide the program's activities to meet the goals, priorities, and anticipated outcomes of the participating agencies. In addition, the act requires the triennial report to address how the program intends to move results out of the laboratory and into application for the benefit of society, its plan for long-term funding for interdisciplinary R&D, and the allocation of funding for interagency projects. The 2007 strategic plan is the first to follow external assessments by the National

Academies and PCAST (operating as the NNAP) and seeks to incorporate the findings of these reviews. Of particular note:

- The 2007 strategic plan includes a new chapter on "High-Impact Application Opportunities and Critical Research Needs" possibly indicating an effort on the part of the Administration to move the NNI toward more directed research with commercial and societal benefits. Much of the early NNI work has been focused on basic research and mechanisms by which such research may produce economic and societal dividends. Seven years into the NNI, the program is under increasing scrutiny to deliver the promised benefits. While the plan does not establish R&D or application targets per se, this chapter illustrates tangible benefits that may be achieved by research supported under the NNI. In addition, the plan provides a matrix that identifies which agencies have a central or supporting role in each key application area. The plan also provides a series of nanotechnology application-specific vignettes on topics such as early detection of life-threatening disease, smarter computers, more energy-efficient transportation, and energy security.
- The 2007 strategic plan split the Societal Dimensions PCA into two separate PCAs: Environmental, Health, and Safety, and Education and Societal Dimensions. This change responds to increased Congressional and public attention to EHS needs. Some critics of the NNI had raised concerns that the inclusion of investments in education and other societal dimensions in the broader category obscured and artificially inflated the perception of investments in EHS R&D.
- The 2007 strategic plan also identifies four areas of common interest across agencies that is to be the focus of future workshops: sensors and nanoelectronics, energy, fate and transport of nanomaterials, and medical and health applications.

Strategy for Nanotechnology-related Environmental, Health, and Safety Research[72]

Strategy for Nanotechnology-related Environmental, Health, and Safety Research outlines the NNI's strategy for addressing nanotechnology EHS concerns. The report outlines the process for developing a research strategy, including identification of priority needs, assessment of existing research, and analysis of strengths and weaknesses of the current portfolio of EHS-focused research. The report provides a summary and analysis of FY2006 EHS research projects using the five priority categories identified in the earlier report, *The National Nanotechnology Initiative: Environmental, Health, and Safety Research Needs for Engineered Nanoscale Materials*[73] (discussed below). The report concludes with a framework for addressing EHS research needs, including a summary and prioritization of research needs, an implementation strategy, and a discussion of interagency coordination efforts.

Prioritization of Environmental, Health, and Safety Research Needs for Engineered Nanoscale Materials: An Interim Document for Public Comment[74]

This document is a follow-on to the *Environmental, Health, and Safety Research Needs for Engineered Nanoscale Materials* report (discussed below), incorporating public comments, refinements of the prioritization principles, and continued assessment of research needs. This chapter further identifies and defines five priorities within each of the five general categories of research needs established in the earlier document and presents the revised principles and the process used for this prioritization. The NEHI working group of the NSET subcommittee expects to use this chapter to evaluate the NNI's current EHS research portfolio, perform a gap analysis, and identify opportunities for interagency collaboration. The report stresses that the NSET subcommittee is "pursuing a dynamic, open, and transparent process in developing an NNI EHS research strategy" and invites continuing public input.

The National Nanotechnology Initiative: Environmental, Health, and Safety Research Needs for Engineered Nanoscale Materials[75]

This chapter documents the efforts of the NSET subcommittee's Nanotechnology Environmental and Health Implications (NEHI) working group to identify, prioritize, and implement research and other activities required for the responsible research and development of nanotechnology. The report is designed to help inform the research, risk assessment, and risk management activities of federal agencies and the private sector.

The report identifies priority research within five general research areas: instrumentation, metrology, and analytical methods; nanomaterials and human health; nanomaterials and the environment; health and environmental surveillance; and risk management methods.

The report identifies several next steps:

- prioritize research needs among those identified in the report;
- evaluate in greater detail the current NNI EHS research portfolio;
- perform a "gap analysis" of the NNI EHS research compared to the prioritized needs;
- coordinate and facilitate among the NNI agencies' research programs to address priorities; and
- establish a process for periodic review of progress and for updating research needs and priorities.

The report concludes that conducting EHS research in parallel with the development of nanomaterials and their applications will help to ensure the full, safe, and responsible realization of the promise of nanotechnology, and that coordination of research activities among NNIparticipating agencies, as well as with industry and other governments, is necessary to expedite progress. In contrast, some NGOs have asserted the need for EHS research to precede the development of nanomaterials and nanotechnology applications.

Selected NNI Assessments

The 21st Century Nanotechnology R&D Act of 2003 (P.L. 108-153) requires periodic external reviews of the National Nanotechnology Program (NNP) by the National Research Council, an arm of the National Academies,[76] and the National Nanotechnology Advisory Panel (NNAP).[77] In general, these reviews concluded that the NNI has been successful so far and that its efforts are important to future U.S. technological leadership and commercial competitiveness. Both reports emphasize that much nanotechnology research is still in its very early stages and caution against expecting too much in the near term from this nascent technology. The reports also laud the cooperative efforts between the NNI and stakeholders in academia and industry and encourage increased interactions with industry, state and local economic developers, and, where appropriate, international partners.

The National Nanotechnology Initiative: Second Assessment and Recommendations of the National Nanotechnology Advisory Panel, President's Council of Advisors on Science and Technology[78]

This chapter presents the findings of the second review of the NNI by the President's Council of Advisors on Science and Technology's, acting as the National Nanotechnology Advisory Panel, as mandated by the 21st Century Nanotechnology R&D Act of 2003. PCAST submitted its first report to the President on May 16, 2005, titled *The National Nanotechnology Initiative at Five Years: Assessment and Recommendations of the National Nanotechnology Advisory Panel.*

As in earlier reports by PCAST and the National Research Council, this chapter concludes that the United States remains a leader in nanotechnology, though it recognizes the successful development of other nations' nanotechnology research capabilities. The PCAST concluded that the NNI's approach to addressing EHS research was "sound" and found proposals for a separate agency or office devoted to nanotechnology EHS research or to set aside a particular percentage of NNI funding for EHS research to be "misguided" and potentially counterproductive as resources may be directed away from research "on beneficial applications and on risk." The panel also concluded that nanotechnology does not raise ethical concerns unique from those accompanying other technological advances. The PCAST also concluded that the NNI is a "highly successful model for an interagency program" describing it as "well organized and well managed."

Recommendations

The report made six broad recommendations for further strengthening the NNI, supported by more specific actions. They are:

PCAST recommendation:

> Maintaining the world-class R&D infrastructure and strong interagency coordination created under the NNI is essential to achieving broad societal benefits from nanotechnology innovation.

Some specific PCAST recommendations include continuing support for NNI multidisciplinary centers, networks and user facilities; improving coordination within agencies that have several operating divisions; and strengthening the participation of the Departments of Commerce, Education, and Labor to address education, training, market assessment, and standards development challenges related to nanotechnology.

PCAST recommendation:

> Progress across the breadth of NNI-supported R&D critically depends upon the development and implementation of standards for nanomaterial identification, characterization, and risk assessment.

In this regard, PCAST specifically recommended that federal agencies participate in the development of voluntary consensus-based standards; develop materials and analytical standards for nanotechnology EHS research; and work towards development of data sets of physical and chemical properties of nanomaterials.

PCAST recommendation:

> Nanotechnology innovation through to commercialization depends on maintaining and strengthening cross-sector collaborations and cross-fertilization of technology development and business development expertise.

Specifically, PCAST recommended expansion of efforts to assess national and international innovation and commercialization activities led by the Department of Commerce; expanded partnering between NNI multidisciplinary centers and economic development organizations; and educated more nanotechnology scientists and engineers to become entrepreneurs and skilled technology workers.

PCAST recommendation:

> Nanotechnology research must be strategically guided, integrated, and coordinated across agencies, sectors, and countries, and include balanced assessment of risks and benefits in the context of specific, real-world applications.

Among the specific PCAST recommendations: improving coordination of federal EHS research efforts with industry and international stakeholders; preventing the segregation of research focused on EHS implications of nanotechnology from research focused on applications; continuing interagency collaboration on EHS research and mission-focused research; and distributing nonproprietary information about the properties of nanomaterials.

PCAST recommendation:

> Research on the societal and ethical aspects of nanotechnology should both be integrated with technical R&D and take place in the context of broader societal and ethical leadership.

PCAST recommendation:

> Public perception of and expectations related to nanotechnology should be informed based on sound science and balanced assessment of risks and benefits (known and anticipated) of specific innovations and their implications for society.

PCAST recommended that the NNI more clearly demonstrate to the public the value of nanotechnology and NNI-supported R&D, calling for a more explicit and direct outreach approach to better inform and engage policymakers, stakeholders, and the general public. In addition PCAST recommended that NNI agencies should provided additional funding to the NNCO to support coordination among NNI agencies to enhance their agency-specific communication efforts.

Addendum to the National Nanotechnology Initiative: Second Assessment and Recommendations of the National Nanotechnology Advisory Panel, President's Council of Advisors on Science and Technology[79]

PCAST submitted an addendum to its second assessment of the NNI in July 2008, providing an assessment of *The National Nanotechnology Initiative Strategy for Nanotechnology-Related Environmental, Health, and Safety Research*. The PCAST letter reaffirmed its judgment (as stated in its second assessment of the NNI) of the NNI's approach to EHS research as "fundamentally sound in terms of strategic priorities as well as focus on and extent of collaboration across agencies, with industry, and particularly with international stakeholders." PCAST found the NNI's strategic approach to be "quite thorough and robust," and stated that the NSET subcommittee's NEHI working group

> remains the best locus of coordination and authoritative advisory capability for the participating agencies in implementing the government-wide nanotechnology EHS research strategy that cuts across agency and disciplinary lines.[80]

PCAST made six recommendations with respect to the NNI's EHS research strategy. First, PCAST recommended that an assessment of the federal nanotechnology EHS portfolio and gap analysis be performed once every three years noting that such a timetable would support the triennial reviews of PCAST and the NRC. Second, it proposed increased collaboration with industry and other countries to fill identified research gaps. Third, PCAST recommended that the NNI encourage its supported researchers to report on analytical methodologies used in their research to facilitate development of best practices for risk assessment and characterization. Fourth, PCAST stated that the NNI should promote broad and practical use of EHS findings in defining responsible use of nanotechnology in research, manufacturing, and commercial applications. Fifth, PCAST called for increased funding for exposure assessment in the context of manufacturing and disposal of nanomaterials and products incorporating relevant quantities of nanomaterials. Sixth, PCAST recommended that the NNI maintain and strengthen agency support and coordination efforts through the NSET subcommittee and the NEHI working group, specifically calling for

> all agencies that fund or conduct research on nanomaterials, as well as those agencies with charters that specifically include EHS-related activities, to directly support the NSET/NEHI data gathering and communications functions.

A Matter of Size: Triennial Review of the National Nanotechnology Initiative[81]

This 2006 report presents the findings of the National Research Council's (NRC) first triennial review mandated by the 21st Century Nanotechnology R&D Act of 2003.

The NRC study concluded that the NNI has been successful in coordinating nanoscale efforts and interests across the federal government, in catalyzing cooperative R&D across a variety of scientific and engineering disciplines, and in opening a host of new scientific opportunities through its infrastructure and R&D investments. The NRC attributed much of this success to effective communication and coordination by the NSET subcommittee and the NNCO.

Recommendations

Here are the recommendations made by the NRC followed by a discussion of each.

NRC recommendation:

> the federal government [should] sustain investments in a manner that balances the pursuit of shorter-term goals with support for longer-term R&D and that ensures a robust supporting infrastructure, broadly defined. Supporting long-term research effectively will require making new funds available that do not come at the expense of much-needed ongoing investment in U.S. physical sciences and engineering research.

President Bush has expressed support for increasing federal R&D funding for the physical sciences and engineering, most notably in his American Competitiveness Initiative which includes nanotechnology investments. Yet, as Federal non-discretionary spending growth increases pressure on federal discretionary spending, finding new funds to support long-term nanotechnology research may need to come from other scientific disciplines.

NRC recommendation:

> the federal government [should] establish an independent advisory panel with specific operational expertise in nanoscale science and engineering; management of research centers, facilities, and partnerships; and interdisciplinary collaboration to facilitate cutting-edge research on and effective and responsible development of nanotechnology.

In July 2004, President George W. Bush implemented the provision of the 21st Century Nanotechnology Research and Development Act to "establish or designate a National Nanotechnology Advisory Panel" by issuing Executive Order 13349, which amends Executive Order 13226, designating the President's Council of Advisors on Science and Technology to serve as the NNAP.

The NRC's recommendation suggests that the President's designation of PCAST to serve as the legislatively mandated National Nanotechnology Advisory Panel is not fully adequate. Critics of the use of PCAST to serve as the NNAP maintain that the scope and depth of expertise needed to provide effective guidance on the NNI requires an independent panel of people with nanotechnology- and interdisciplinary-specific expertise and an undivided focus. Supporters of the use of PCAST for this function assert that a single advisory panel provides an integrated perspective, reduces unnecessary cost and management burdens, and that expertise can be added to the panel or accessed through non-member technical advisory groups.

NRC recommendation:

> federal agencies participating in the NNI, in consultation with the NNCO and the Office of Management and Budget, should continue to develop and enhance means for consistent tracking and reporting of funds requested, authorized, and expended annually. The current set of PCAs provides an appropriate initial template for such tracking.

It is difficult to assess and track funding for specific purposes within the NNI because the initiative is not centrally funded and operated. The NNI budget is an aggregation of the nanotechnology-related activities of the participating federal agencies. Congress funds the

NNIrelated R&D on an agency-by-agency basis, with responsibilities crossing many authorizing committees and appropriations subcommittees. Thus, while it is relatively straightforward to quantify an agency's nanotechnology budget, tracking all NNI investments related to a particular activity—EHS-related research, for example—is much more difficult. The PCAs serve to provide such a tracking mechanism. In addition, according to the 2007 *National Nanotechnology Initiative Strategic Plan*, the division of the Societal Dimensions PCA into two PCAs— Environmental, Health, and Safety; and Educational and Societal Dimensions—is intended to better understand and manage the NNI investment. Such a change indicates a level of flexibility that may enable the executive branch and Congress to more effectively manage and balance investments in discrete areas of the NNI.

NRC recommendation:

> the NSET Subcommittee [should] carry out or commission a study on the feasibility of developing metrics to quantify the return to the U.S. economy from the federal investment in nanotechnology R&D. The study should draw on the Department of Commerce's expertise in economic analysis and its existing ability to poll U.S. industry. Among the activities for which metrics should be developed and relevant data collected are technology transfer and commercial development of nanotechnology.

Few efforts have been made within the federal government to understand the economic impacts of the nation's investments in the NNI. Identification and tracking of data that could serve as an indicator of success in commercializing nanotechnology research or the effects on U.S. job creation or retention has not been formalized. To the extent that federal assessments of the economic contribution of and/or potential for nanotechnology products have occurred, they have not been performed with analytical rigor. Although the Commerce Department retains its economic analysis expertise, resident primarily in the Economics and Statistics Administration's Bureau of Economic Analysis, the Department's Technology Administration, which led Commerce's NNI activities and had government-wide responsibilities for technology transfer activities, was eliminated in August 2007.[82] Prior to its elimination, the Technology Administration contracted for two studies that could contribute to addressing this NRC recommendation: an analysis of barriers to nanotechnology commercialization performed by the University of Illinois at Springfield, and an analysis of innovation metrics conducted by the Alliance for Science and Technology Research in America (ASTRA). These reports are publicly available at Commerce Department websites.[83]

NRC recommendation:

> research on the environmental, health, and safety effects of nanotechnology [should] be expanded. Assessing the effects of engineered nanomaterials on public health and the environment requires that the research conducted be well-defined and reproducible, and that effective methods be developed and applied to (1) estimate the exposure of humans, wildlife, and other ecological receptors to source material; (2) assess effects on human health and ecosystems of both occupational and environmental exposure; and (3) characterize, assess, and manage the risks associated with exposure.

While the NRC asserts the need for additional EHS research, it does not quantify how much more is needed. Clayton Teague, director of the NNCO, has testified that the current level of investment in EHS research is adequate.[84] Many critics from academia, industry, and non-profit organizations have argued strongly that the NNI needs a greater level of investment in EHS research.[85] These critics argue from a variety of perspectives, including the need to:

- protect workers, human health, and the environment;
- create public faith and confidence in the safety of nanotechnology products;
- prevent a problem with one specific nanotechnology product from resulting in a loss of public support for all nanotechnology R&D; and
- create a predictable and stable regulatory environment.

This last factor is deemed by some as critical to fostering future nanotechnology investments.

NRC recommendation:

> the NSET Subcommittee [should] create a working group on education and the workforce that engages the Department of Education and Department of Labor as active participants.

The NSET subcommittee has sought, with limited success, greater involvement of the Departments of Education and Labor in the subcommittee's activities. An NSET subcommittee working group on education and the workforce has not yet been established.

With advocates promising the creation of many new jobs—some assert millions—as a result of global nanotechnology investments, some have expressed concern that the country must prepare students for nanotechnology research, engineering, and production jobs.[86] Assessing which industries are likely to create such jobs, which skills will be needed, and in what timeframe are key challenges. If workers with nanotechnology-specific skills are needed and no workers are available domestically (U.S. citizens, resident aliens, or those in the United States on work visas), potential employers may opt to establish or move operations outside the United States to tap workers with those skills abroad. Conversely, if students are trained for jobs that do not emerge or do not emerge in the same timeframe as students are entering the job market, this investment is lost. In addition, potential students may be discouraged from pursuing future nanotechnologyrelated studies. Close coordination among the Departments of Commerce, Education, and Labor might help to align federal education and training efforts better with the labor market for nanotechnology workers.

The 21st Century Nanotechnology R&D Act also directed the NRC to address two other issues in its first triennial report: Is molecular self assembly feasible for manufacturing of materials and devices at the nanoscale? And, what are the needs for standards, guidelines, or strategies for ensuring the responsible development of nanotechnology?

Molecular Self-Assembly

Self-assembly is the process by which components (atoms, molecules, or more complex structures) form, without external control or direction, an organized structure. For example, water molecules dispersed in air in cold temperatures can self-assemble to form snowflakes. Our bodies act as self-assemblers, producing a variety of cells as needed (e.g. to repair a damage to the skin or produce new blood cells from added nutrients).

To what extent can molecular self-assembly be used as a tool for nanomanufacturing? On this issue, the NRC concluded that molecular self-assembly is feasible for the manufacture of simple materials and devices. However, for the manufacture of more sophisticated materials and devices, including complex objects produced in large quantities, the NRC found it unlikely that simple self-assembly processes will yield the desired results. One major barrier cited is the probability of error during assembly as a result of the systems' complexity.

Standards, Guidelines, and Strategies for Ensuring Responsible Development of Nanotechnology

The NRC concluded that it is not possible yet to make a rigorous assessment of the level of environmental and health risks posed by engineered nanomaterials and called for further development of risk assessment protocols. The NRC report also stated that the need for more EHS data requires an expanded research effort to complement dialog on these issues. In addition, until reproducible and well-characterized EHS data are available to inform the development of rigorous risk-based guidelines and best practices, the NRC found it prudent to recommend use of precautionary measures to protect the health and safety of workers, the public, and the environment. The NRC report also stressed that addressing the ethical and societal impacts of nanotechnology will require an integrated approach among scientists, engineers, social scientists, toxicologists, policymakers, and the public.

The National Nanotechnology Initiative at Five Years: Assessment and Recommendations of the National Nanotechnology Advisory Panel, President's Council of Advisors on Science and Technology[87]

This chapter presents the findings of the first biennial review of the NNI by the President's Council of Advisors on Science and Technology's, acting as the National Nanotechnology Advisory Panel, as mandated by the 21st Century Nanotechnology R&D Act of 2003. The PCAST report finds that the United States is the acknowledged leader in nanotechnology R&D, but the U.S. leadership position is under increasing competitive pressure from growing public and private investments around the world. The report states that the federal investment in the NNI has been well-spent, the United States is well-positioned to maintain global leadership going forward, and continued robust funding is important for long-term U.S. economic well-being and national security. This assessment of the U.S. leadership position is founded not on sales, growth, or market share of commercial products—common measures of global competitiveness for established products— but rather on metrics that may serve as early indicators of potential innovation, such as the U.S. share of scientific publications and patents. The use of such metrics may not be universally accepted as predictive of leadership position. Technological leadership—or even leadership in innovation—does not ensure that the economic benefits from such leadership will accrue to the United States. Companies may choose to manufacture products or conduct other value-added activities outside the United States. If the assessment of national competitiveness is expanded to include the value-added activities and jobs generated or retained within the United States, then the metrics for assessing leadership might change.

The PCAST report acknowledges that there are potential environmental and health risks associated with nanotechnology, but finds that the NNI is directing appropriate attention and adequate resources to the research that will ensure the protection of the public and the environment. Nanotechnology products should not be immune from regulation, according to the report, but such regulation must be rational and based on science, not on perceived fears.

The PCAST report states that strong communication exists among the NNI agencies responsible for research and regulation. The PCAST report contains four recommendations for the NNI:

PCAST recommendation:

> To further facilitate technology transfer from the lab to the marketplace, the NNI should expand its interaction with industry, increase federal-state coordination, and improve knowledge management of and access to NNI assets, such as user facilities and instrumentation.

The NSET subcommittee's National Innovation and Liaison with Industry (NILI) working group was established to facilitate NNI interactions with industry, and with state and local nanotechnology initiatives. The NILI working group's limited resources and agency participation have hindered its ability to conduct more extensive and sustained outreach.[88] Due to the structure and resource allocation of the NNI, the initiative's engagements with industry and with state and local initiatives are largely limited to single agency or laboratory interactions and to public engagement activities, such as speeches and information on the NNI website.[89]

PCAST recommendation:

> The NNI should continue its efforts to understand the possible toxicological effects of nanotechnology and where harmful human or environmental effects are proven, pertinent federal agencies should apply appropriate regulatory mechanisms. There should be strong interagency and international collaboration on this issue to eliminate unnecessary duplication of research efforts and to ensure wide dissemination of information. Since exposure to nanomaterials is most likely to occur during the manufacturing process, research on potential hazards associated with workplace exposure must be given the highest priority.

With respect to collaboration on EHS issues, the NSET subcommittee's National Environmental and Health Implications (NEHI) working group is the primary EHS coordination mechanism for participating NNI agencies. The Global Issues in Nanotechnology (GIN) working group works with the NEHI working group on international collaboration on EHS issues. The NIOSH has published several documents addressing concerns about workplace exposure to nanoparticles. *Approaches to Safe Nanotechnology: An Information Exchange With NIOSH* was intended to provide the best currently available knowledge on nanoparticle toxicity and control and to solicit input from the stakeholder community.[90] *Progress Toward Safe Nanotechnology in the Workplace* details the work of NIOSH's Nanotechnology Research Center from 2004 through 2006.[91] In December 2007, NIOSH released interim guidance concerning the medical screening of workers potentially exposed to engineered nanoparticles during the manufacture and industrial use of nanomaterials. The NIOSH says that the document is intended to "generate discussion, fill the current knowledge gap, and provide interim recommendations until further scientific information becomes available."[92] The NIOSH is currently seeking public comment on this guidance.

With respect to regulatory issues associated with nanotechnology, see CRS Report RL34332, *Engineered Nanoscale Materials and Derivative Products: Regulatory Challenges*, by Linda-Jo Schierow.

PCAST recommendation:

> The NNI should establish relationships with the Department of Education and Department of Labor to develop education and training systems to support the Nation's technical proficiency in areas related to nanotechnology.

The PCAST report's recommendation is similar to the recommendation made by the NRC and is discussed earlier in this paper.

PCAST recommendation:

> The NNI must support research aimed at understanding the societal implications of nanotechnology—including ethical, economic and legal implications—and must actively work to inform the public about nanotechnology. The NNI should continue to confront societal issues in an open, straightforward, and science-based manner.

Some critics of the NNI hold deep reservations about the ethical, economic, and legal implications of nanotechnology. Some of these concerns are common to many technologies, such as the allocation of risk and benefit during manufacturing. For example, a neighborhood located near a production facility may bear risks associated with exposure to the byproducts (or products) of manufacturing, while gaining few of the benefits. Concerns about possible adverse effects of nanoscale particles on human health and the environment resulting from their small particle size and unique characteristics may result in increased attention to such costs and benefits with respect to nanoscale material production. Currently, nanotechnology EHS risks are unknown and may be acute or pose no more risk than other manufacturing processes.

Privacy rights are another issue associated with the products of nanotechnology. Nanotechnology may enable the production of highly sensitive, inexpensive sensors that could be deployed ubiquitously in commercial and public settings. While these sensors may allow check-out-free purchases from stores, or monitor the environment for toxic substances, critics argue that they could also impinge on the privacy rights of individuals if, for example, the sensors could detect chemicals related to the use of tobacco, alcohol, or illegal substances without the permission of the individual. Such information might be later applied in law enforcement, life insurance, health insurance, or employment decisions.[93] Others express concern that the economically disadvantaged and less educated—both individuals and nations—might be unable to take part in the benefits that nanotechnology products could offer.[94]

On the legal front, innovations in nanotechnology are already presenting unique challenges to the U.S. Patent and Trademark (USPTO). For example, U.S. case law generally prohibits patenting where the sole element of novelty is a change in size, the characteristic most obviously associated with nanotechnology.[95] In addition, many nanotechnology innovations involve multiple disciplines. Since the USPTO structure for examining patents is discipline-based, an examiner may not have all of the requisite expertise for the examination, affecting both their ability to conduct the examination, and the speed at which it can be done.

USPTO also has acknowledged the need to accelerate the speed of nanotechnology-related patent applications. According to John Doll, Commissioner of Patents, the agency is hampered in its ability to recruit and retain patent examiners with the requisite skills to handle nanotechnology patents given the "more generous offers [patent examiners get] from the private sector."[96] Doll said that efforts have been made to improve hiring and retention at USPTO, and that a new processes has been established allowing an accelerated examination of applications.[97]

SELECTED NANOTECHNOLOGY LEGISLATION IN THE 111TH CONGRESS

Title I, Subtitle A, H.R. 5116—National Nanotechnology Initiative Amendments Act of 2010

The provisions of Title I, Subtitle A of H.R. 5116, the National Nanotechnology Initiative Amendments Act of 2010, are nearly identical to H.R. 554 (see "H.R. 554—National Nanotechnology Initiative Amendments Act of 2009" below). H.R. 5116 changes the name of the act from the "National Nanotechnology Initiative Amendments Act of 2009," to "National Nanotechnology Initiative Amendments Act of 2010," and removes the term "interdisciplinary" from a provision establishing "green nanotechnology" research centers.

H.R. 554—National Nanotechnology Initiative Amendments Act of 2009

H.R. 554, the National Nanotechnology Initiative Amendments Act of 2009, was introduced on January 15, 2009, and passed by the House of Representatives on February 11, 2009. The bill was referred to the Senate Commerce, Science, and Transportation Committee on February 12, 2009. The purpose of the bill is to authorize activities for support of nanotechnology research and development and for other purposes. Among its provisions, the bill would amend the 21st Century Nanotechnology Research and Development Act of 2003 to:

- require the NSTC triennial strategic plan to include near-term and long-term objectives, the anticipated timeframe for achieving near-term objectives, and metrics for assessing progress; cooperative and collaborative activities in R&D and technology transition supported by the states; and proposed research in areas of national priority;
- require the NSTC annual nanotechnology report supplementing the President's budget request to include a breakout of spending for the development and acquisition of research facilities and instrumentation for each program component area, and a breakout of spending on all activities related to ethical, legal, environmental, and societal implications;
- direct NNP agencies to support the activities of committees involved in the development of standards for nanotechnology and allow agencies to reimburse the travel costs of scientists and engineers who participate in activities of such committees;

- direct the agencies to fund the National Nanotechnology Coordination Office, and to do so in proportion to each agency's share of the previous year's NNP budget;
- require the NNCO to develop and maintain a publicly accessible database of projects funded under the Environmental, Health, and Safety, the Education and Societal Dimensions, and the Nanomanufacturing program component areas;
- require the NNCO to develop, maintain, and publicize information on nanotechnology facilities supported by the NNP, including at a minimum the terms and conditions for the use of each facility, a description of the capabilities of the instruments and equipment available for use at the facility, and a description of the technical support available to assist users of the facility;
- require the establishment of a National Nanotechnology Advisory Panel (NNAP) "as a distinct entity." Currently, under the provisions of presidential Executive Order 13349, the President's Council of Advisors on Science and Technology serves as the NNAP;[98]
- direct the NNCO to enter into an arrangement with the National Research Council to conduct a triennial review of the NNP, and authorizes funds for FY2010, FY2011, and FY2012; and
- define nanotechnology as "the science and technology that will enable one to understand, measure, manipulate, and manufacture at the nanoscale, aimed at creating materials, devices, and systems with fundamentally new properties or functions," and define nanoscale as "one or more dimensions of between approximately 1 and 100 nanometers."

In addition, the bill would:

- require the designation of a White House Office of Science and Technology Policy associate director to serve as the "Coordinator for Societal Dimensions of Nanotechnology" and would charge the coordinator with convening and chairing a panel of federal agency representatives and others to develop, maintain, implement, and monitor an annual EHS research plan that includes, among other things, standards related to nanotechnology nomenclature; standards for methods and procedures for detecting, measuring, monitoring, sampling, and testing engineered nanoscale materials for environmental, health, and safety impacts; and standard reference materials for EHS testing;
- require the National Science Foundation to provide grants to establish Nanotechnology Education Partnerships to recruit and help prepare secondary school students to pursue postsecondary level courses of instruction in nanotechnology;
- direct the NSTC to establish an Education Working Group under the NSET Subcommittee to coordinate, prioritize, and plan NNP educational activities;
- direct certain NNP agencies to provide companies access to their supported facilities to assist in the development of prototypes of nanoscale products, devices, or processes for determining proof of concept;
- direct NNP agencies to encourage nanotechnology-related submissions to their Small Business Innovation Research (SBIR) and Small Business Technology Transfer (STTR) programs;
- direct NIST to encourage nanotechnology-related submissions to its Technology Innovation Program (TIP), and directs the TIP advisory Board to provide advice to NIST to

accomplish this, and to provide an assessment of the adequacy of TIP resources allocated to nanotechnology related projects;

- direct the NSTC to actively pursue industry liaison groups for all industries;
- direct the NNP to coordinate and leverage federal investments with nanotechnology research, development, and technology transition initiatives supported by the States;
- direct the NNP to support nanotechnology R&D activities directed toward application areas that have the potential for significant contributions to national economic competitiveness and for other significant societal benefits, such as nano-electronics, energy efficiency, health care, and water remediation and purification;
- direct the NNP to support research on the development of instrumentation and tools required for the rapid characterization of nanoscale materials and for monitoring of nanoscale manufacturing processes, and to support approaches and techniques for scaling the synthesis of new nanoscale materials to achieve industrial-level production rates; and
- direct certain NNP-supported interdisciplinary research centers to support research on methods and approaches to environmentally benign nanoscale products and nanoscale manufacturing processes, as well as related technology transfer and education activities.

S. 1482—National Nanotechnology Amendments Act of 2009

S. 1482, the National Nanotechnology Amendments Act of 2009, was introduced on July 21, 2009, and referred to the Senate Commerce, Science, and Transportation Committee. The purpose of the bill is to reauthorize the 21st Century Nanotechnology Research and Development Act and to expand the scope of the National Nanotechnology Program (NNP).

Among its provisions, the bill:

- requires the NNP to solicit and draw upon the perspectives of the industrial community to promote the rapid commercial development of nanoscale-enabled devices, systems, and technologies and to coordinate research in determining the key physical and chemical characteristics of nanoparticles and nanomaterials that may pose environmental, health, and safety risks;
- requires the NNCO and other appropriate agencies and councils to issue guidance to agencies that describes a strategy for transitioning research into commercial products and technologies and how the program will coordinate or conduct research on the environmental, health, and safety issues related to nanotechnology;
- requires the NSTC triennial strategic plan to include near-term and long-term objectives, the anticipated timeframe for achieving near-term objectives, and metrics for assessing progress; cooperative and collaborative activities in R&D and technology transition supported by the states; how the NNP intends to encourage and support interdisciplinary research; and proposed research in areas of national priority;
- encourages joint interagency solicitation of grant applications in high priority, multi-disciplinary research areas;
- requires participating agencies to support the activities of the committees of standards setting bodies involved in the development of standards for nanotechnology;

- requires each participating agency to provide funds to support the work of the NNCO. Authorizes appropriations to: (1) NIST for the development of nanotechnology standards; and (2) NSF, for use by the NNCO, to develop and maintain a public information database of NNP projects in EHS; education; public outreach; ethical, legal, and other societal issues; and of nanotechnology facilities accessible for use by individuals from academia and industry;
- makes the National Nanotechnology Advisory Panel (NNAP) a distinct entity, and requires the NNAP to establish a subpanel to enable it to assess whether societal, ethical, legal, environmental, and workforce concerns are adequately addressed by the NNP;
- revises provisions for triennial external review of the NNP;
- requires the designation of a "coordinator for societal dimensions of nanotechnology," within OSTP, to convene a panel to develop a research plan, and requires the coordinator to enter into an arrangement with the National Science Board to create a report that identifies the broad goals and needs of EHS researchers;
- directs the NSTC to establish an interagency Education Working Group to coordinate, prioritize, and plan formal and informal educational activities supported under the NNP, including activities to help participants understand the EHS implications of nanotechnology;
- provides for one or more grants to establish Nanotechnology Education Partnerships to recruit and help prepare secondary school students to pursue postsecondary level courses in nanotechnology;
- requires agencies supporting nanotechnology research facilities to provide access to representatives from industry and other stakeholders for the transfer of research results or assist in developing proof-of-concept prototypes of nanoscale products, devices, or processes;
- directs NIST, in its Technology Innovation Program, and all agencies with Small Business Innovation Research (SBIR) and Small Business Technology Transfer (STTR) programs, to encourage the submission of nanotechnology related grant proposals;
- sets, for the NNP, the objective of establishing industry liaison groups for all industry sectors that would benefit from nanotechnology applications;
- requires coordination and leveraging of federal investments with nanotechnology research, development, and technology transition initiatives supported by state governments;
- requires the NNP to support nanotechnology R&D in areas of national importance (e.g., economic competitiveness, energy production, water purification, agriculture, and health care; in environmental, health, and safety research on the risks of nanoparticles) and in ethical, legal, and societal issues related to nanotechnology;
- requires the NNP to support a wide array of research in support of nanomanufacturing;
- requires the director of the NNCO to review and report on nanomanufacturing research and research facilities;
- requires an NNAP review of the nanomanufacturing program component area and the capabilities of nanotechnology research facilities supported by the NNP;
- sets forth provisions regarding NNP nanoscale characterization and metrology research; and

- requires deliberative public input in the decision making processes affecting policies for the research, development, and use of nanotechnology, and authorizes $2.0 million for the NNCO to carry out this responsibility.

S. 596—Nanotechnology Innovation and Prize Competition Act of 2009

S. 596, the Nanotechnology Innovation and Prize Competition Act of 2009, was introduced on March 16, 2009, and referred to the Senate Commerce, Science, and Transportation Committee. The purpose of the bill is to establish an award program to honor achievements in nanotechnology. Under the bill, the Department of Commerce's National Institute of Standards and Technology is directed to award prizes to individuals and companies for achievement in one or more of the following areas: improvement of the environment, consistent with EPA's Twelve Principles of Green Chemistry; development of alternative energy that has the potential to lessen the dependence of the United States on fossil fuels; and/or improvement of human health, consistent with regulations promulgated by the FDA. The bill authorizes financial prizes for being the first to achieve a specific criteria, as well as recognition prizes, made as part of the previously established National Technology and Innovation Medal award program. The bill authorizes $2,000,000 annually for the financial prizes as well as $750,000 annually for administration of the prize competitions.

H.R. 820—Nanotechnology Advancement and New Opportunities Act

H.R. 820, the Nanotechnology Advancement and New Opportunities Act, was introduced on February 3, 2009, and referred to four House committees: the Committee on Science and Technology, the Committee on Energy and Commerce, the Committee on Ways and Means, and the Committee on Homeland Security. The purpose of the bill is to ensure the development and responsible stewardship of nanotechnology. The bill would:

- establish a $100 million Nanomanufacturing Investment Partnership at the Department of Commerce to work with private investors to advance the commercialization of nanomanufacturing technologies and to increase the commercial application of federally supported research results;
- establish a 15% tax credit, taken over five years, for the purchase of up to $10 million of stock in qualified nanotechnology companies; and
- establish a grant program within the DOC to support the establishment and development of nanotechnology incubators by non-profit entities and degree- granting institutions;
- require the Director of the NNCO to prepare a report to Congress on a nanotechnology research strategy for government and industry that will ensure the development and responsible stewardship of nanotechnology;
- provide a tax credit of 50% for nanotechnology education and training expenses for businesses and individuals;

- authorize an annual appropriation of $15 million for FY20 10 through FY20 13 for the NSF to conduct a grant program for the development of curriculum materials for interdisciplinary nanotechnology courses at institutions of higher education;
- direct the NSF to establish, through its Advanced Technological Education program, a program to encourage manufacturing companies to enter into partnerships with occupational training centers for the development of training to support nanomanufacturing; and
- direct the Secretary of Energy to submit a report to Congress containing a strategy for increasing interaction among scientists and engineers at DOE national laboratories and the informal science education community to prepare appropriate exhibits for school age children and the general public.

In addition, the bill would amend the 21st Century Nanotechnology Research and Development Act of 2003 to:

- authorize $10 million for the NSF to establish a center for the development of computer-aided design tools for nanotechnology applications;
- authorize an annual appropriation of $30 million for the DOE to conduct a grant program for nanotechnology research to address the need for "clean, cheap, renewable energy;"
- authorize an annual appropriation of $30 million for the EPA for a grant program for nanotechnology research to address technologies for the remediation of pollution and other environmental protection technologies;
- authorize an annual appropriation of $30 million for the DHS to conduct a grant program for nanotechnology research to address the need for sensors and materials related to homeland security needs; and
- authorize an annual appropriation of $30 million for the DHHS to conduct a grant program for nanotechnology research to address health-related applications.

H.R. 2647—National Defense Authorization Act for Fiscal Year 2010

Section 242 of the National Defense Authorization Act for Fiscal Year 2010 (H.R. 2647, P.L. 111- 84) amends the Department of Defense's nanotechnology reporting responsibilities to align with those required of other agencies under the 21st Century Nanotechnology Research and Development Act (P.L. 108-153). H.R. 2647 was signed into law on October 28, 2009.

S. 3117—Promote Nanotechnology in Schools Act

S. 3117, the Promote Nanotechnology in Schools Act, was introduced on March 15, 2010, and referred to the Senate Committee on Health, Education, Labor, and Pensions. The purpose of the bill is to strengthen the capacity of eligible institutions (i.e., secondary schools, community colleges, two-year and four-year institutions of higher education, and informal learning science centers) to provide instruction in nanotechnology. The bill would authorize a program at the National Science Foundation for this purpose that would offer eligible institutions grants of up to $400,000 (subject to a 25% match from non-federal sources) to assist in the purchase and maintenance of nanotechnology equipment and software, to develop and provide educational services, and to support teacher education and certification. The bill would authorize $15 million for FY20 10 and "such sums as may be necessary" for FY20 11 through FY 2013.

H.R. 4502—Nanotechnology Education Act

H.R. 4502, the Nanotechnology Education Act, was introduced on February 19, 2010, and referred to the House Committee on Science and Technology's Subcommittee on Research and Science Education. The purpose of the bill is to strengthen the capacity of eligible institutions (i.e., secondary schools, community colleges, four-year institutions of higher education, and informal learning science centers) to provide instruction in nanotechnology. The bill would authorize a program for this purpose at the National Science Foundation that would offer eligible institutions grants of up to $400,000 (subject to a 25% match from non-federal sources) to assist in the purchase and maintenance of nanotechnology equipment and software, to develop and provide educational services, and to support teacher education and certification. The bill would authorize $40 million for FY20 11 and "such sums as may be necessary" for FY20 12 through FY 2014.

S. 2942—Nanotechnology Safety Act of 2010

S. 2942, the Nanotechnology Safety Act of 2010, was introduced on January 21, 2010, and referred to the Senate Committee on Health, Education, Labor, and Pensions. The bill would require the Secretary of Health and Human Services to establish within 180 days a program for the scientific investigation of nanoscale materials included or intended for inclusion in FDA- regulated products, to address the potential toxicology of such materials, the effects of such materials on biological systems, and interaction of such materials with biological systems. The bill would authorize $25 million per year for fiscal years 2011 to 2015.

CONCLUDING OBSERVATIONS

Many expect nanotechnology to bring significant economic and societal returns. The United States was the first government to launch a national-level nanotechnology program and has invested more than any other nation. As a result of this focus and these sustained investments, many experts believe that the United States enjoys a technological leadership position in nanotechnology. Other nations are investing heavily and some industrialized and emerging economies have formidable capabilities in nanotechnology. Assessments of the National Nanotechnology Initiative have concluded that the effort is well-managed and has been successful in achieving its objectives so far. However, these assessments have recognized that the NNI faces a variety of challenges in ensuring that the full promise of nanotechnology is realized and that the United States remains the global leader in nanoscale science, engineering, and technology.

Congress may choose to address some or many of the issues addressed in the body of this chapter in the course of deliberation on the reauthorization of the 21st Century Nanotechnology R&D Act of 2003 or, alternatively, in separate legislation.

The 21st Century Nanotechnology R&D Act's funding authorizations extended through FY2008. Action is being considered in both the House and Senate on reauthorization of the program. Possible topics for consideration in the reauthorization process include budget authorization levels for the covered agencies; R&D funding levels, priorities, and balance across the program component areas; administration and management of the NNI; translation of research results and early-stage technology into commercially viable applications; environmental, health, and safety issues; ethical, legal, and societal implications; education and training for the nanotechnology workforce; metrology, standards, and nomenclature; public understanding; and international dimensions. Consideration may also be given to the establishment of an independent review panel and to coordination of the timing for the NNAP assessment, the NRC assessment, and the NSET subcommittee's strategic plan for the NNI.

APPENDIX A. SELECTED REPORTS ON THE NATIONAL NANOTECHNOLOGY INITIATIVE

Reports of the Nanoscale Science, Engineering, and Technology Subcommittee of the National Science and Technology Council

The National Nanotechnology Initiative: Research and Development Leading to a Revolution in Technology and Industry, Supplement to the President's FY2010 Budget. May 2009.

The National Nanotechnology Initiative: Research and Development Leading to a Revolution in Technology and Industry, Supplement to the President's FY2009 Budget. September 2008.

National Nanotechnology Initiative Fiscal Year 2009 Budget Summary & Highlights. February 2008

Strategy for Nanotechnology-Related Environmental, Health, and Safety Research. February 2008

The National Nanotechnology Initiative Strategic Plan. December 2007.

The National Nanotechnology Initiative: Research and Development Leading to a Revolution in Technology and Industry, Supplement to the President's FY2008 Budget. July 2007.

The National Nanotechnology Initiative: Environmental, Health, and Safety Research Needs for Engineered Nanoscale Materials. September 2006.

The National Nanotechnology Initiative: Research and Development Leading to a Revolution in Technology and Industry, Supplement to the President's FY2007 Budget. July 2006.

The National Nanotechnology Initiative: Research and Development Leading to a Revolution in Technology and Industry, Supplement to the President's FY2006 Budget. March 2005.

The National Nanotechnology Initiative Strategic Plan. December 2004.

Nanotechnology in Space Exploration. August 2004.

Nanoscience Research for Energy Needs. Report from a workshop held in March 2004.

Instrumentation and Metrology for Nanotechnology. Report from a workshop held in January 2004.

Nanotechnology: Societal Implications-Maximizing Benefits for Humanity. Report from a workshop held in December 2003.

Nanobiotechnology. Report from a workshop held in October 2003.

Regional, State, and Local Initiatives in Nanotechnology. Report from a workshop held in September-October 2003.

National Nanotechnology Initiative: Research and Development Supporting the Next Industrial Revolution, Supplement to the President's FY2004 Budget. August 2003.

Nanotechnology and the Environment. Report from a workshop held in May 2003.

National Nanotechnology Initiative: The Initiative and Its Implementation Plan, Detailed Technical Report Associated with the Supplemental Report to the President's FY2003 Budget. June 2002.

National Nanotechnology Initiative: The Initiative and Its Implementation Plan, Detailed Technical Report Associated with the Supplemental Report to the President's FY2001 Budget. July 2000.

Report of the Interagency Working Group on Nanoscience, Technology, and Engineering (NSET Subcommittee Predecessor)

Nanotechnology: Shaping the World Atom by Atom. 1999.

Agency Reports

NIOSH Update: NIOSH Draft Offers Interim Guidance on Medical Screening of Workers Potentially Exposed to Engineered Nanoparticles, National Institute for Occupational Safety and Health. December 2007.

Progress Toward Safe Nanotechnology in the Workplace, National Institute for Occupational Safety and Health. June 2007.

Approaches to Safe Nanotechnology in the Workplace, National Institute for Occupational Safety and Health. July 2006.

Nanoscale Science, Engineering, and Technology in DOE's Office of Basic Energy Sciences, U.S. Department of Energy. February 2003.

External Reviews

A Matter of Size: Triennial Review of the National Nanotechnology Initiative, National Research Council. 2006.

The National Nanotechnology Initiative at Five Years: Assessment and Recommendations of the National Nanotechnology Advisory Panel, President's Council of Advisors on Science and Technology (acting as the National Nanotechnology Advisory Panel). May 2005.

Small Wonders, Endless Frontiers: A Review of the National Nanotechnology Initiative, National Research Council. June 2002.

APPENDIX B. LIST OF NNI AND NANOTECHNOLOGY-RELATED ACRONYMS

ASTRA	Alliance for Science and Technology Research in America
CNST	Center for Nanoscale Science and Technology
CS	Committee on Science
CT	Committee on Technology
CSREES	Cooperative State Research, Education, and Extension Service
DHHS	Department of Health and Human Services
DHS	Department of Homeland Security
DOC	Department of Commerce
DOD	Department of Defense
DOE	Department of Energy
DOJ	Department of Justice
DOT	Department of Transportation
EHS	Environmental, health, and safety
ELSI	Ethical, legal, and societal implications
EPA	Environmental Protection Agency
EOP	Executive Office of the President
EPSCoR	Experimental Program to Stimulate Competitive Research
FHWA	Federal Highway Administration
GIN	Global Issues in Nanotechnology working group
ISO	International Standards Organization
IWGN	Interagency Working Group on Nanotechnology
NASA	National Aeronautics and Space Administration
NCI	National Cancer Institute
NEHI	National Environmental and Health Implications working group
NGO	Non-governmental organization
NIH	National Institutes of Health
NILI	National Innovation and Liaison with Industry working group
NIOSH	National Institute of Occupational Safety and Health
NIST	National Institute of Standards and Technology
NNAP	National Nanotechnology Advisory Panel

NNCO	National Nanotechnology Coordination Office
NNI	National Nanotechnology Initiative
NNIN	National Nanotechnology Infrastructure Network
NNN	National Nanomanufacturing Network
NNP	National Nanotechnology Program
NPEC	Nanotechnology Public Engagement and Communications working group
NRC	National Research Council
NSET	Nanoscale Science, Engineering, and Technology subcommittee
NSF	National Science Foundation
NSEC	Nanoscale Science and Engineering Center
NSRC	Nanoscale Science Research Centers
NSTC	National Science and Technology Council
OECD	Organization for Economic Cooperation and Development
OMB	Office of Management and Budget
OSTP	Office of Science and Technology Policy
PCA	Program Component Areas
PCAST	President's Council of Advisors on Science and Technology
R&D	Research and development
SBIR	Small Business Innovation Research
STTR	Small Business Technology Transfer Research
TSA	Transportation Security Administration
USDA	U.S. Department of Agriculture
USPTO	U.S. Patent and Trademark Office

End Notes

[1] President Obama requested $1.6 billion in nanotechnology R&D funding for FY2010. Although Congress has passed all of the regular appropriations acts, the precise amount provided for FY20 10 is currently under evaluation by the agencies and the White House Office of Management and Budget and will likely be published with the President's FY20 11 budget request in February 2010.

[2] In the context of the NNI and nanotechnology, the nanoscale refers to a dimension of 1 to 100 nanometers (see box on this page).

[3] While extensive R&D has been, and continues to be, conducted to understand and harness the properties of individual atoms, this is not the domain of nanotechnology.

[4] *Nanotechnology Takes a Deep Breath... and Prepares to Save the World,* Cientifica, April 2009, available at http://www.cientifica.eu/files/Whitepapers/Nanotechnology%20Takes%20a%20Deep%20Breath.pdf.

[5] The National Nanotechnology Initiative at Five Years: Assessment and Recommendations of the National Nanotechnology Advisory Panel, President's Council of Advisors on Science and Technology, May 2005, available at http://www.nano.gov/html/res/FINAL_PCAST_NANO_REPORT.pdf.

[6] *Nanotechnology Takes a Deep Breath... and Prepares to Save the World,* Cientifica, April 2009, available at http://www.cientifica.eu/files/Whitepapers/Nanotechnology%20Takes%20a%20Deep%20Breath.pdf.

[7] Ibid.

[8] "Overhyped Technology Starts to Reach Potential," press release, Lux Research, July 22, 2008. http://www.luxresearchinc.com/press/RELEASE_Nano-SMR_7_22_08.pdf. Numbers do not add to total due to rounding.

[9] Ibid

[10] *Halfway to the Trillion Dollar Market: A Critical Review of the Diffusion of Nanotechnologies*, Cientifica, 2007. http://www.cientifica.eu/files/Whitepapers/A%20Reassessment%20of%20the%20Trillion%20WP.pdf

[11] While views vary on how to calculate nanotechnology's contribution to these products, the consensus is that nanotechnology is likely to have a significant economic impact and transformative effect on many industries.

[12] National Cancer Institute website. http://nano.cancer.gov/resource_center/tech_backgrounder.asp

[13] Ibid.

[14] Aslan, Kadir; Lakowicz, Joseph R.; and Geddes, Chris D. "Nanogold plasmon resonance-based glucose sensing. Wavelength-ratiometric resonance light scattering," *Analytical Chemistry*, 2005, Vol. 77. *Strategic Plan for Pediatric Urology,* National Institute of Diabetes and Digestive and Kidney Disease, National Institutes of Health, Department of Health and Human Services, February 2006.

[15] Risbud, Aditi. "Fruit of the Nano Loom," *Technology Review*, February 2006.

[16] *Nanoscience Research for Energy Needs,* Nanoscale Science, Engineering, and Technology Subcommittee, National Science and Technology Council, The White House, December 2004.

[17] Risbud, Aditi. "Cheap Drinking Water from the Ocean," *Technology Review*, June 2006.

[18] Selko, Adrienne. "New Nanotechnology-Based Coatings Are Energy Efficient and Environmentally Sound," *Industry Week*, August 22, 2007. "Tomorrow's Green Nanofactories," *Science Daily*, July 11, 2007.

[19] *National Nanotechnology Initiative—Leading to the Next Industrial Revolution,* Interagency Working Group on Nanoscience, Engineering, and Technology, National Science and Technology Council, The White House. http://www.ostp.gov/NSTC/html/iwgn/iwgn.fy01budsuppl/nni.pdf

[20] *21st Century Agriculture: A Critical Role for Science and Technology*, U.S. Department of Agriculture, June 2003; and *Nanoscale Science and Engineering for Agriculture and Food Systems*: *Draft Report of the National Planning Workshop to the Cooperative State Research, Education, and Extension Service of the U.S. Department of Agriculture*, July 2003.

[21] *Nanotechnology in Space Exploration*, Nanoscale Science, Engineering, and Technology Subcommittee, National Science and Technology Council, The White House, August 2004, available at http://www.nano.gov/nni_space_exploration_rpt.pdf.

[22] *Nanotechnology and the Environment*, Nanoscale Science, Engineering, and Technology Subcommittee, National Science and Technology Council, The White House, May 2003, available at http://www.nano.gov/NNI_Nanotechnology_and_the_Environment.pdf.

[23] *Proceedings of the U.S. Environmental Protection Agency Workshop on Nanotechnology for Site Remediation,* U.S. Environmental Protection Agency, October 2005.

[24] "National Nanotechnology Initiative: Leading to the Next Industrial Revolution," press release, The White House, January 21, 2000. http://clinton4.nara.gov/WH/New/html/20000121_4.html; and "Steering the technology that will redefine life as we know it," *Industrial Biotechnology*, Vol. 1, No. 3, Fall 2005. http://www.nsf.gov/crssprgm/nano/ reports/mcr_ind_biotech_interview.pdf

[25] This figure includes $ 1.657 billion in regular appropriations and $0. 140 billion in funding provided by the American Recovery and Reinvestment Act of 2009 (P.L. 111-5).

[26] While many provisions of this act have no sunset provision, FY2008 was the last year of agency authorizations included in the act.

[27] Prior to P.L. 108-153, the Bob Stump Defense Authorization Act for Fiscal Year 2003 (P.L. 107-314) required DOD to "provide for interagency cooperation and collaboration on nanoscale research and development." The NSET subcommittee is a subcommittee of the NSTC Committee on Technology.

[28] Metrology is the science of measurement, including the equipment and processes used to produce a measurement.

[29] The Department of Commerce has characterized emerging and enabling technologies as those that "offer a wide breadth of potential application and form an important technical basis for future commercial applications." (ATP Rule, 15 C.F.R. Part 295).

[30] The term "valley of death" is used by business executives, economists, and venture capitalists to describe the development gap that often exists between a laboratory discovery and the market's willingness to invest to advance the discovery to a final commercial product. This gap occurs due to a variety of issues, such as technical risk, market uncertainty, and likelihood of obtaining an adequate return on investment.

[31] Crews, Clyde Wayne, Jr., "Washington's Big Little Pork Barrel: Nanotechnology," Cato Institute website, May 29, 2003.

[32] Testimony of Andrew Maynard, Chief Science Advisor, Project on Emerging Nanotechnologies, Woodrow Wilson International Center for Scholars, "Research on Environmental and Safety Impacts of Nanotechnology: Current Status of Planning and Implementation under the National Nanotechnology Initiative," hearing, Subcommittee on Research and Science Education, House Committee on Science and Technology, October 31, 2007.

[33] "NGOs urge precautionary principle in use of nanomaterials," EurActiv.com, June 14, 2007. http://www.euractiv.com/en/environment/ngos-urge-precautionary-principle-use-nanomaterials/article-164619 Sass, Jennifer. "Nanotechnology and the Precautionary Principle," presentation, Natural Resources Defense Council, 2006. http://docs.nrdc.org/health/hea_06121402a.pdf The precautionary principle has been used in other countries on some issues. For example, the Biosafety Protocol to the 1992 Convention on Biological Diversity incorporates provisions applying the precautionary principle to the safe handling, transfer, and trade of genetically modified organisms. For further information, see CRS Report RL3 0594, *Biosafety Protocol for Genetically Modified Organisms: Overview*, by Alejandro E. Segarra and Susan R. Fletcher.

[34] "No Small Matter II: The Case for a Global Moratorium—Size Matters!," Occasional Paper Series, ETC Group, April 2003. http://www.etcgroup.org/upload/publication/pdf_file/165

[35] *The National Nanotechnology Initiative Strategic Plan,* Nanoscale Science, Engineering, and Technology Subcommittee, National Science and Technology Council, The White House, December 2004, available at http://www.nano.gov/NNI_Strategic_Plan_2004.pdf.

[36] Ibid.

[37] National Nanotechnology Initiative website. http://www.nano.gov/html/about/history.html

[38] "National Nanotechnology Initiative: Leading to the Next Industrial Revolution," press release, The White House, January 21, 2000. http://clinton4.nara.gov/WH/New/html/20000121_4.html; and National Nanotechnology Initiative website. http://www.nano.gov/html/about/history.html

[39] In its January 21, 2001 press release, "National Nanotechnology Initiative: Leading to the Next Industrial Revolution," announcing the establishment of the NNI, the White House identified only six participating agencies— NSF, DOD, DOE, NIST, NASA, and NIH. Subsequently, EPA and DOJ reported nanotechnology R&D funding in FY2001, bringing the total number of agencies funding nanotechnology R&D in FY2001 to eight.

[40] NNI participants include agencies that either conduct or provide funding for nanotechnology R&D, as well as agencies with missions that may affect the development, commercialization, and use of nanotechnology. For example, in the latter case, the Food and Drug Administration may regulate (or not regulate) nanotechnology products, the U.S. Patent and Trademark Office's (USPTO) treatment of nanotechnology-related patents may affect the value of the underlying intellectual property, and the execution of the missions of the Departments of Education and Labor could affect the preparedness of the U.S. workforce for emerging nanotechnology jobs. Some nanotechnology R&D agencies may also have non-R&D missions related to nanotechnology. For example, EPA conducts and funds R&D but also has a regulatory mission that could affect nanotechnology research, development, production, use, and/or disposal.

[41] The original six agencies identified at the launch of the NNI were the Department of Defense, Department of Energy, National Institute of Standards and Technology (Department of Commerce), National Science Foundation, National Aeronautics and Space Administration, and National Institutes of Health (DHHS). "National Nanotechnology Initiative: Leading to the Next Industrial Revolution," press release, The White House, January 21, 2000. http://clinton4.nara.gov/WH/New/html/20000121_4.html; and National Nanotechnology Initiative website. http://www.nano.gov/html/about/history.html

[42] U.S. Congress. 2003. 21st Century Nanotechnology Research and Development Act. P.L. 108-153. 15 U.S.C. 7501. 108 Cong., December 3.

[43] FY2003 funding attributed to DHS for the purpose of this calculation is based on nanotechnology R&D appropriations received by the Department of Transportation's Transportation Security Administration (TSA). TSA was transferred to DHS in the Homeland Security Act of 2002 (P.L. 107-296) which was enacted after the start of FY2003.

[44] *The National Nanotechnology Initiative Strategic Plan*, Nanoscale Science, Engineering, and Technology Subcommittee, National Science and Technology Council, The White House, December 2004, available at http://www.nano.gov/NNI_Strategic_Plan_2004.pdf.

[45] The agencies that participate in the NSET subcommittee comprise the NNI. NSET subcommittee members include Bureau of Industry and Security, Department of Commerce; Consumer Product Safety Commission; National Institute of Food and Agriculture, Department of Agriculture; Department of Defense; Department of Education; Department of Energy; Department of Homeland Security; Department of Justice; Department of Labor; Department of State; Department of Transportation; Department of the Treasury; Environmental Protection Agency; Food and Drug Administration; Forest Service, Department of Agriculture; Director of National Intelligence; International Trade Commission; National Aeronautics and Space Administration; National Institutes of Health, U.S. Department of Health and Human Services; National Institute for Occupational Safety and Health, Center for Disease Control, Department of Health and Human Services; National Institute of Standards and Technology, Department of Commerce; National Science Foundation; Nuclear Regulatory Commission; U.S. Geological Survey; and U.S. Patent and Trademark Office, Department of Commerce. The Department of Commerce's Technology Administration was a participating agency in the NNI until its elimination in August 2007 under the America COMPETES Act (P.L. 110-69).

[46] The National Nanotechnology Initiative: Research and Development Leading to a Revolution in Technology and Industry-Supplement to the President's FY2008 Budget, Nanoscale Science, Engineering, and Technology Subcommittee, National Science and Technology Council, The White House, July 2007, available at http://www.nano.gov/NNI_08Budget.pdf.

[47] Testimony of E. Clayton Teague, Director, National Nanotechnology Coordinating Office, *Research on Environmental and Safety Impacts of Nanotechnology: Current Status of Planning and Implementation under the National Nanotechnology Initiative*, hearing, Subcommittee on Research and Science Education, House Committee on Science and Technology, October 31, 2007.

[48] The Industrial Research Institute is an association of companies and federally funded laboratories with the mission of improving R&D capabilities through the development and dissemination of best practices.

[49] National Nanotechnology Initiative website, http://www.nano.gov.

[50] The eight agencies that are signatories to the memorandum of understanding are NSF, DOD, DOE, NIH, NIST, NASA, EPA, and DOT.

[51] This figure does not include NNI funding provided by ARRA. Agency ARRA allocations included $5 11.2 million for nanotechnology R&D. Agencies may report additional ARRA funding for SBIR and STTR projects later, when 2009 SBIR/STTR data become available.

[52] *The National Nanotechnology Initiative Strategic Plan,* Nanoscale Science, Engineering, and Technology Subcommittee, National Science and Technology Council, The White House, December 2004, http://www.nano.gov/ NNI_Strategic_Plan_2004.pdf.

[53] *The National Nanotechnology Initiative Strategic Plan,* Nanoscale Science, Engineering, and Technology Subcommittee, National Science and Technology Council, The White House, December 2007, available at http://www.nano.gov/NNI_Strategic_Plan_2007.pdf.

[54] *The National Nanotechnology Initiative: FY2010 Budget & Highlights*, Nanoscale Science, Engineering, and Technology Subcommittee, National Science and Technology Council, The White House, May 2009, available at http://www.nano.gov/NNI_2010_budget_supplement.pdf.

[55] Ibid.

[56] Ibid.

[57] Ibid.

[58] Top-down processes are those that achieve design features by removing material from a larger block of material; bottom-up processes begin with smaller building blocks (atoms or molecules) and achieve design features by putting them together, possibly using self-assembly.

[59] *The National Nanotechnology Initiative: FY2010 Budget & Highlights*, Nanoscale Science, Engineering, and Technology Subcommittee, National Science and Technology Council, The White House, May 2009, available at http://www.nano.gov/NNI_2010_budget_supplement.pdf.

[60] Ibid.

[61] *The National Nanotechnology Initiative: FY2010 Budget & Highlights*, Nanoscale Science, Engineering, and Technology Subcommittee, National Science and Technology Council, The White House, May 2009, available at http://www.nano.gov/NNI_2010_budget_supplement.pdf.

[62] Ibid.

[63] *The National Nanotechnology Initiative: FY2010 Budget & Highlights*, Nanoscale Science, Engineering, and Technology Subcommittee, National Science and Technology Council, The White House, May 2009, available at http://www.nano.gov/NNI_2010_budget_supplement.pdf.

[64] Ibid.

[65] In addition, 18 other Materials Research Science and Engineering Centers conduct nanotechnology-related research as part of their overall efforts.

[66] Private telephone communication between CRS and NSTC staff, January 31, 2008.

[67] Private e-mail communication between CRS and NSF staff, January 31, 2008.

[68] Private telephone communication between CRS and NSTC staff, January 31, 2008.

[69] Private e-mail communication between CRS and NSF staff, January 31, 2008.

[70] *The National Nanotechnology Initiative: FY2011 Budget & Highlights*, Nanoscale Science, Engineering, and Technology Subcommittee, National Science and Technology Council, The White House, February 2010, available at http://www.nano.gov/NNI_2011_budget_supplement.pdf.

[71] *The National Nanotechnology Strategic Plan*, Nanoscale Science, Engineering, and Technology Subcommittee, National Science and Technology Council, The White House, December 2007, available at http://www.nano.gov/ NNI_Strategic_Plan_2007.pdf.

[72] *Strategy for Nanotechnology-related Environmental, Health, and Safety Research*, Science, Engineering, and Technology Subcommittee, National Science and Technology Council, The White House, February 2008, available at http://www.nano.gov/NNI_EHS_Research_Strategy.pdf.

[73] *The National Nanotechnology Initiative: Environmental, Health, and Safety Research Needs for Engineered Nanoscale Materials*, Nanoscale Science, Engineering, and Technology Subcommittee, National Science and Technology Council, The White House, September 2006, available at http://www.nano.gov/ NNI_EHS_research_needs.pdf.

[74] Prioritization of Environmental, Health, and Safety Research Needs for Engineered Nanoscale Materials: An Interim Document for Public Comment, Nanoscale Science, Engineering, and Technology Subcommittee, National Science and Technology Council, The White House, August 2007, available at http://www.nano.gov/ Prioritization_EHS_Research_Needs_Engineered_Nanoscale_Materials.pdf.

[75] *The National Nanotechnology Initiative: Environmental, Health, and Safety Research Needs for Engineered Nanoscale Materials*, Nanoscale Science, Engineering, and Technology Subcommittee, National Science and Technology Council, The White House, September 2006, available at http://www.nano.gov/ NNI_EHS_research_needs.pdf.

[76] The National Research Council, the National Academy of Sciences, the National Academy of Engineering, and the Institute of Medicine are part of a private, non-profit institution established under a congressional charter. They are collectively referred to as the National Academies.

[77] P.L. 108-153 directs the President to "establish or designate a National Nanotechnology Advisory Panel." In July 2004, President Bush designated the President's Council of Advisors on Science and Technology to serve as the NNAP by issuing Executive Order 13349, *Amending Executive Order 13226 To Designate the President's Council of Advisors on Science and Technology To Serve as the National Nanotechnology Advisory Panel.*

[78] The National Nanotechnology Initiative: Second Assessment and Recommendations of the National Nanotechnology Advisory Panel, President's Council of Advisors on Science and Technology, May 2005, available at http://www.whitehouse.gov/sites/default/files/microsites/ostp/PCAST-NNAP-NNI-Assessment-2008.pdf.

[79] *Addendum to The National Nanotechnology Initiative: Second Assessment and Recommendations of the National Nanotechnology Advisory Panel*, President's Council of Advisors on Science and Technology, May 2005, available at National Research Council, 2006, available at http://www.whitehouse.gov /sites/default/files/microsites/ostp/PCASTAddendum-Letter.pdf.

[80] Ibid.

[81] *A Matter of Size: Triennial Review of the National Nanotechnology Initiative*, National Research Council, 2006, available at http://books.nap.edu/catalog.php?record_id=11752.

[82] The Technology Administration was eliminated in the America COMPETES Act (P.L. 110-69).

[83] *Barriers to Nanotechnology Commercialization*, College of Business Management, University of Illinois at pringfield, September 2007, available at http://www.osec.doc.gov/ReportBarriers%20to %20Nanotechnology%20Commercialization.pdf *Innovation Vital Signs Project*, Alliance for Science and Technology Research in America, July 2007. http://www.ntis.gov/ta_reports/Report-InnovationVitalSigns.pdf.

[84] Testimony of Clayton Teague, director of the NNCO, hearing, "Environmental and Safety Impacts of Nanotechnology: What Research is Needed?" House Committee on Science, November 17, 2005. http://commdocs.house.gov/committees/science/hsy24464.000/hsy24464_0.HTM

[85] Ibid.

[86] Phillip J. Bond, Under Secretary for Technology, U.S. Department of Commerce, remarks, "Nanotechnology: Economic Opportunities, Societal and Ethical Challenges," NanoCommerce 2003, December 9, 2003. http://www.technology.gov/Speeches/PJB_031209.htm *Sizing Nanotechnology 's Value Chain*, Lux Research, October 2004.

[87] The National Nanotechnology Initiative at Five Years: Assessment and Recommendations of the National Nanotechnology Advisory Panel, President's Council of Advisors on Science and Technology, May 2005, available at http://www.nano.gov/html/res/FINAL_PCAST_NANO_REPORT.pdf.

[88] Private telephone and e-mail communication with Sean Murdock, executive director of the NanoBusiness Alliance, February 4, 2008.

[89] Ibid.

[90] *Approaches to Safe Nanotechnology: An Information Exchange with NIOSH*, National Institute for Occupational Safety and Health, July 2006.

[91] *Progress Toward Safe Nanotechnology in the Workplace*, National Institute for Occupational Safety and Health, June 2007.

[92] *NIOSH Update: NIOSH Draft Offers Interim Guidance on Medical Screening of Workers Potentially Exposed to Engineered Nanoparticles*, National Institute for Occupational Safety and Health, December 13, 2007.

[93] Moore, Fiona M., "Implications of Nanotechnology Applications: Using Genetics as a Lesson," *Health Law Review*, Vol. 10, No. 3, 2002. http://www.law.ualberta.ca/centres/hli/pdfs/hlr/v10_3/10.3moorefrm.pdf

[94] Smith, Richard H. ,"Social, Ethical, and Legal Implications of Nanotechnology," *Societal Implications of Nanoscience and Nanotechnology* (The Netherlands :Kluwer Academic Publishers, 2001).

[95] *Nanotechnology Patents: Issues for Nanotechnology Inventions*, Dorsey and Whitney, LLP, May 9, 2005.

[96] Doll, John, Commissioner of Patents, U.S. Patent and Trademark Office, Letter to the Editor, *Small Times*, April 23, 2007. http://www.smalltimes.com/Articles/Article_Display.cfm?ARTICLE_ID=290818&p=109

[97] Ibid.

[98] Executive Order 13349 is available at http://edocket.access.gpo.gov/cfr_2005/janqtr/3CFR13349.htm.

In: Nanotechnology: Looking Ahead
Editor: Julian S. Moser

ISBN: 978-1-61761-785-0

Chapter 3

The National Nanotechnology Initiative: Research and Development Leading to a Revolution in Technology and Industry

National Science and Technology Council

1. Introduction and Overview

Overview of the National Nanotechnology Initiative

The National Nanotechnology Initiative (NNI) is a U.S. Government research and development (R&D) program of twenty-five agencies working together toward the common challenging vision of *"a future in which the ability to understand and control matter at the nanoscale leads to a revolution in technology and industry that benefits society."* The combined, coordinated efforts of these agencies have accelerated discovery, development, and deployment of nanotechnology towards agency missions and the broader national interest. Established in 2001,[1] the NNI involves nanotechnology-related activities by the 25 agencies shown in Table 1, 15 of which have budgets for nanotechnology R&D for 2011.

The National Nanotechnology Initiative is managed within the framework of the National Science and Technology Council (NSTC), the Cabinet-level council by which the President coordinates science and technology policy across the Federal Government. The Nanoscale Science, Engineering, and Technology (NSET) Subcommittee of the NSTC coordinates planning, budgeting, program implementation, and review of the initiative. The NSET Subcommittee is composed of representatives from agencies participating in the NNI. A listing of official NSET Subcommittee members is provided at the front of this chapter. Contact information for NSET Subcommittee participants is provided in Appendix B. The National Nanotechnology Coordination Office (NNCO) acts as the primary point of contact for information on the NNI; provides technical and administrative support to the NSET Subcommittee; supports the subcommittee in the preparation of multiagency planning, budget, and assessment documents, including this chapter; develops, updates, and maintains the NNI website (http://www.nano.gov); and provides public outreach on behalf of the NNI.

The NSET Subcommittee has established four working groups: (1) the Global Issues in Nanotechnology (GIN) Working Group, (2) the Nanotechnology Environmental and Health Implications (NEHI) Working Group, (3) the Nanomanufacturing, Industry Liaison, and Innovation (NILI) Working Group, and (4) the Nanotechnology Public Engagement and Communication (NPEC) Working Group.

The December 2007 NNI Strategic Plan' sets out the vision for the NNI stated above. The plan specifies four goals aimed at achieving that overall vision: (1) advance a world-class nanotechnology research and development program; (2) foster the transfer of new technologies into products for commercial and public benefit; (3) develop and sustain educational resources, a skilled workforce, and the supporting infrastructure and tools to advance nanotechnology; and (4) support responsible development of nanotechnology. The plan also lays out eight NNI investment categories (or program component areas, PCAs), each aimed at helping to achieve one or more of the above goals:

WHAT IS NANOTECHNOLOGY?

Nanotechnology is the understanding and control of matter at dimensions between approximately 1 and 100 nanometers, where unique phenomena enable novel applications. Encompassing nanoscale science, engineering, and technology, nanotechnology involves imaging, measuring, modeling, and manipulating matter at this length scale.

A nanometer is one-billionth of a meter. A sheet of paper is about 100,000 nanometers thick; a single gold atom is about a third of a nanometer in diameter. Dimensions between approximately 1 and 100 nanometers are known as the nanoscale. Unusual physical, chemical, and biological properties can emerge in materials at the nanoscale. These properties may differ in important ways from the properties of bulk materials and single atoms or molecules.

1. Fundamental nanoscale phenomena and processes
2. Nanomaterials
3. Nanoscale devices and systems
4. Instrumentation research, metrology, and standards for nanotechnology
5. Nanomanufacturing
6. Major research facilities and instrumentation acquisition
7. Environment, health, and safety
8. Education and societal dimensions

Federal agencies are investing in R&D within the above categories in support of national goals and agency missions. NNI funding represents the sum of the nanotechnology-related funding allocated by each of the participating agencies. Each agency separately determines its budgets for nanotechnology R&D, in coordination with the Office of Management and Budget, the Office of Science and Technology Policy, and Congress. Thus, the NNI is an interagency budget crosscut in which participating agencies work closely with each other to create an integrated program through communication, coordination, and collaboration.

Enhanced communications through committee and working groups has led to joint coordination and collaboration in a variety of forms, including sharing of knowledge and expertise; joint sponsorship of solicitations and workshops; and leveraging funding, staff, and facility/equipment resources at NNI participating agencies. Examples of focused R&D collaboration across agencies include the Nanotechnology Signature Initiatives summarized in the sidebar on page 6.

Purpose of this Chapter

This document provides supplemental information to the President's 2011 Budget and serves as the Annual Report on the NNI called for in the 21' Century Nanotechnology Research and Development Act (P.L. 108-153, 15 USC 7501). In particular, the report summarizes NNI programmatic activities for 2009 and 2010, as well as those planned for in 2011. This chapter also meets the requirements for the Department of Defense of 10 USC 2358 "Research and Development Projects" as amended by the National Defense Authorization Act for 2010 (H.R. 2647). NNI budgets for 2009-2011 are presented by agency and PCA in Section 2 of this chapter. Information on the use of the Small Business Innovation Research (SBIR) and Small Business Technology Transfer (STTR) program funds to support nanotechnology research and commercialization activities, also called for in P.L. 108-153, is included at the end of Section 2. Section 3 summarizes the perspectives of the NNI agencies regarding their nanotechnology investments. Section 4 highlights activities that have been undertaken and progress that has been made toward achieving the goals set out in the NNI Strategic Plan and discusses external reviews of the NNI and how their recommendations are being addressed.

Table 1. List of Federal Agencies Participating in the NM During 2010

Federal agencies with budgets dedicated to nanotechnology research and development
Consumer Product Safety Commission (CPSC) Department of Defense (DOD) Department of Energy (DOE) Department of Homeland Security (DHS) Department of Justice (DOD Department of Transportation (DOT, including the Federal Highway Administration, FHWA) Environmental Protection Agency (EPA) Food and Drug Administration (FDA, Department of Health and Human Services) Forest Service (FS, Department of Agriculture) National Aeronautics and Space Administration (NASA) National Institute for Occupational Safety and Health (NIOSH, Department of Health and Human Services/Centers for Disease Control and Prevention) National Institute of Food and Agriculture (NIFA, Department of Agriculture) National Institutes of Health (NIH, Department of Health and Human Services)

Table 1. (Continued)

National Institute of Standards and Technology (NIST, Department of Commerce)
National Science Foundation (NSF)
Other participating agencies
Bureau of Industry and Security (BIS, Department of Commerce)
Department of Education (DOEd)
Department of Labor (DOL)
Department of State (DOS)
Department of the Treasury (DOTreas)
Director of National Intelligence (DNI)
International Trade Commission (ITC)
Nuclear Regulatory Commission (NRC)
U.S. Geological Survey (USGS, Department of the Interior)
U.S. Patent and Trademark Office (USPTO, Department of Commerce)

NANOTECHNOLOGY SIGNATURE INITIATIVES

To accelerate nanotechnology development in support of the President's priorities and innovation strategy, NNI member agencies have identified areas ripe for significant advances through close and targeted program-level interagency collaboration. The resulting Nanotechnology Signature Initiatives for 2011 are summarized as follows (details are available at http://www.nano.govihtm Vresea rch/sig nature_i nitiatives.htm I):

Nanotechnology Applications for Solar Energy (DOE, NIST, NSF, DOD, DNI, USDA/NIFA – $51 million total)*

The President's Agenda calls for the development of carbon-neutral alternative energy sources to mitigate global climate change, reduce dependence on foreign oil, improve the economy, and improve the environment. The specific targets specified by the President's Agenda state that 10 percent of electricity generated should be derived from renewable sources by 2012 and 25 percent by 2025.

Solar energy is a promising alternative energy source that can address these challenges. It is readily available, free from geopolitical tension and not a threat to the environment through pollution or to the climate through greenhouse gas emission. The development of a solar energy infrastructure will not only support U.S. energy independence, but also represents an unparalleled economic opportunity if the U.S. can maintain scientific and industrial leadership in this field.

Today, the current rate of improvement of solar technology performance is inadequate to meet future energy needs, and the cost is often not economically competitive without subsidies. Therefore, new innovations and fundamental breakthroughs are needed to accelerate the development of economical solar energy technologies that overcome the limits of existing technologies.

Nanotechnology provides a key to overcoming current performance barriers and substantially improving the collection and conversion of solar energy. At the nanoscale, a number of physical phenomena have been identified that can improve the collection and conversion of solar energy. Nanoparticles and nanostructures have been shown to enhance the absorption of light, increase the conversion of light to electricity, and provide better thermal storage and transport. However, current demonstrations of these technologies fall short of potential performance because of poor control over feature size and placement, unpredictable micro/nanostructure, and poor interface formation. The goal of this initiative is to exploit the benefits of nanotechnology by enhancing understanding of conversion and storage phenomena at the nanoscale, improving nanoscale characterization of electronic properties, and helping enable economical nanomanufacturing.

Sustainable Nanomanufacturing (NIST, NSF, DOE, EPA, NIH – $23 million total)*

A long-term vision for nanomanufacturing is to create flexible, "bottom-up" batch assembly methods that can be used to assemble elaborate systems of complex nanodevices. To create the foundation for achieving this vision, over the next decade the goal of this initiative is to accelerate the development of industrial-scale methods for manufacturing of functional systems with relatively limited complexity based on manufactured nanoparticles with designed properties. The organized assemblies of nanoparticles manufactured will be designed to control and manipulate information, thermal energy, and electromagnetic radiation. The systems to be manufactured, based on these methods, will include disruptive technologies for high-speed communication and computation, solar energy harvesting, waste heat management and recovery, and energy storage. If the initiative is successful, the methods developed will be immediately extendable to more complex components and systems as future nanodevices mature.

Nanoelectronics for 2020 and Beyond (NSF, DOD, NIST, DOE, DNI – $55 million total)*

Continuing to shrink the dimensions of electronic devices is important in order to further increase processing speed, reduce device switching energy, increase system functionality, and reduce manufacturing cost per bit. But as the dimensions of critical elements of devices approach atomic size, quantum tunneling and other quantum effects degrade and ultimately prohibit conventional device operation. Researchers are therefore pursuing somewhat radical approaches to overcome these fundamental physics limitations. Candidate approaches include different types of logic using cellular automata or quantum entanglement and superposition; 3-D spatial architectures; and information-carrying variables other than electron charge, such as photon polarization, electron spin, and position and states of atoms and molecules. Approaches based on nanoscale science, engineering, and technology are the most promising for realizing these radical changes and are expected to change the very nature of electronics and the essence of how electronic devices are manufactured. Rapidly reinforcing domestic R&D successes in these arenas could establish a U.S. domestic manufacturing base that will dominate 21st-century electronics commerce. The goal of this initiative is to accelerate the discovery and use of novel nanoscale fabrication processes and innovative concepts to produce revolutionary materials, devices, systems, and architectures to advance the field of nanoelectronics.

These Nanotechnology Signature Initiatives represent the leading edge of functional interagency collaboration in the budget and program planning process under the NNI, with multiple agencies working in common toward specific objectives.

* Funding commitments by participating agencies at the time this supplement was completed are listed above. Updated/new commitments will be provided at http://www.nano.gov/html/research/signature_initiatives.html.

2. NNI Investments

Budget Summary

The 2011 Budget provides nearly $1.8 billion for the National Nanotechnology Initiative (NNI), reflecting continued steady growth in the NNI investment. This sustained major investment in nanotechnology research and development (R&D) across the Federal Government over the past eleven years of the NNI, spanning three Presidential Administrations and five Congresses, is an indication of the broad support for this program. This support is based on nanotechnology's potential to vastly improve our fundamental understanding and control of matter at the nanoscale, ultimately leading to a revolution in technology and industry for the benefit of society. While the NNI remains focused on fulfilling the Federal role of supporting basic research, infrastructure development, and technology transfer, the proposed investments for 2011 place renewed emphasis on accelerating the transition from basic R&D advances and capabilities into innovations that support national priorities such as sustainable energy technologies, healthcare, and environmental protection. This is consistent with substantial increases in the requested nanotechnology investments for 2011 at the Department of Energy, the National Institutes of Health, and the Environmental Protection Agency. The NNI is also increasing its investments aimed at implementing the Government's strategy for nanotechnology-related environmental, health, and safety (EHS) research.[4] As a part of this expanded EHS effort, the Food and Drug Administration and the Consumer Product Safety Commission are participating in the formal NNI budget crosscut for the first time in 2011. Increasing investments in nanotechnology R&D by other NNI participating agencies reflect the potential for this research to support diverse agency missions and responsibilities. The cumulative NNI investment since 2001, including the 2011 request, now totals over $14 billion. This includes $511 million in funding authorized under the American Recovery and Reinvestment Act of 2009 (P.L. 111-5). Cumulative investments in EHS research since 2005 now total over $480 million. Cumulative investments in education and in research on ethical, legal, and other societal dimensions of nanotechnology since 2005 total over $260 million.

The 2011 NNI budget supports nanoscale science, engineering, and technology R&D at 15 agencies. Agencies with the largest investments are:

- DOE (research providing a basis for new and improved energy technologies)
- NSF (fundamental research across all disciplines of science and engineering)
- NIH (nanotechnology-based biomedical research at the intersection of life sciences and the physical sciences)
- DOD (science and engineering rcsearch advancing defense and dual-use capabilities)

- NIST (fundamental research and development of measurement and fabrication tools, analytical methodologies, and metrology for nanotechnology)

Table 2. NNI Budget, 1 by Agency, 2009-2011 (dollars in millions)

Agency	2009 Actual	2009 Recovery*	2010 Estimated	2011 Proposed
DOE**	332.6	293.2	372.9	423.9
NSF	408.6	101.2	417.7	401.3
HHS/NIH	342.8	73.4	360.6	382.4
DOD***	459.0	0.0	436.4	348.5
DOC/NIST	93.4	43.4	114.4	108.0
EPA	11.6	0.0	17.7	20.0
HHS/NIOSH	6.7	0.0	9.5	16.5
NASA	13.7	0.0	13.7	15.8
HHS/FDA	6.5	0.0	7.3	15.0
DHS	9.1	0.0	11.7	11.7
USDA/NIFA	9.9	0.0	10.4	8.9
USDA/FS	5.4	0.0	5.4	5.4
CPSC	0.2	0.0	0.2	2.2
DOT/FHWA	0.9	0.0	3.2	2.0
DOJ	1.2	0.0	0.0	0.0
TOTAL****	**1,701.5**	**511.3**	**1,781.1**	**1,761.6**

*Based on allocations of the American Recovery and Reinvestment Act of 2009 (P.L. 111-5) appropriations. Agencies may report additional ARRA funding for SBIR and STTR projects later, when 2009 SBIR/STTR data become available.

** Funding levels for DOE include the Office of Science, the Office of Energy Efficiency and Renewable Energy, the Office of Fossil Energy, the Office of Nuclear Energy, and the Advanced Research Projects Agency-Energy.

*** In Tables 2-4, the 2009 and 2010 DOD figures include Congressionally directed funding that is outside the NNI plan ($117 million for 2009).

****For Tables 2-7, totals may not add, due to rounding.

Other agencies investing in mission-related research are EPA, NIOSH, NASA, FDA, DHS, USDA (including both NIFA and FS), CPSC, DOT (including FHWA), and DOJ.

Table 2 shows NNI investments in 2009-2011 for Federal agencies with budgets and investments for nanotechnology R&D Tables 3-6 list the investments for 2009-2011 by agency and by program component area (PCA).

Key points about the 2010 and 2011 NNI investments

- Beyond the $1.7 billion in total NNI investments reported under the respective 2009 agency appropriations, an additional $511 million was provided for nanotechnology research and infrastructure investments in 2009 through the American Recovery and Reinvestment Act (ARRA) of 2009. This substantial ARRA enhancement to the NNI in 2009 included investments across all of the NNI PCAs, with particular emphasis on PCA 2 (nanomaterials, where DOE alone increased its investment by $138 million) and on PCA 6 (major research facilities & instrumentation acquisition), where NIST, DOE, and NSF allocated over $72 million to upgrade equipment and facilities at NNI user facilities and research centers.

Table 3. Actual 2009 Agency Investments by Program Component Area (Doller in Millions)

	1. Fundametal Phenomena & Processes	2. Nanometerials	3. Nanoscale Devices & Systems	4. Instrument Research Metrology & Standards	5. Nano-manufacturing	6. Major research Facilities & Instr. Acquisition	7. Environment, Health, and Safety	8. Education & Societal Dimensions	NNI Toal
DOE	99.8	92.8	7.9	21.1	6.9	100.5	3.1	0.5	332.6
NSF	143.6	72.4	54.0	21.4	27.7	31.5	26.8	31.3	408.6
HHS/NIH	46.7	73.9	172.4	17.5	2.2	13.6	12.0	4.5	342.8
DOD	162.8	67.5	166.9	9.0	29.0	19.7	4.1	0.0	459.0
DOC/NIST	23.1	8.5	17.0	19.4	9.4	12.4	3.5	0.0	93.4
EPA	0.2	0.2	0.1	0.0	0.0	0.0	11.1	0.0	11.6
HHS/NIOSH	0.0	0.0	0.0	0.0	0.0	0.0	6.7	0.0	6.7
NASA	0.0	8.6	5.1	0.0	0.0	0.0	0.0	0.0	13.7
HHS/FDA	0.0	0.0	0.0	0.0	0.0	0.0	6.5	0.0	6.5
DHS	0.0	3.7	5.3	0.0	0.1	0.0	0.0	0.0	9.1
USDA/NIFA	1.0	2.0	5.7	0.0	0.2	0.0	0.5	0.5	9.9
USDA/FS	2.0	1.4	0.7	1.1	0.2	0.0	0.0	0.0	5.4
CPSC	0.0	0.0	0.0	0.0	0.0	0.0	0.2	0.0	0.2
DOT/FHWA	0.0	0.9	0.0	0.0	0.0	0.0	0.0	0.0	0.9
DOJ	0.0	0.0	0.0	1.2	0.0	0.0	0.0	0.0	1.2
TOTAL	**479.2**	**331.9**	**435.2**	**90.8**	**75.6**	**177.6**	**74.5**	**36.8**	**1701.5**

Table 4. 2009 Agency Investments from the ARRA by Program Component Area (Dollars in Millions)

	1. Fundametal Phenomena & Processes	2. Nanometerials	3. Nanoscale Devices & Systems	4. Instrument Research Metrology & Standards	5. Nano-manufacturing	6. Major research Facilities & Instr. Acquisition	7. Environment, Health, and Safety	8. Education & Societal Dimensions	NNI Toal
DOE	85.9	137.9	21.8	3.5	18.9	25.0	0.2	0.0	293.2
NSF	29.9	24.7	17.6	4.5	6.1	6.5	3.4	8.5	101.2
HHS/NIH	14.7	15.7	28.6	4.4	0.4	0.7	8.4	0.5	73.4
DOC/NIST	0.0	0.0	0.0	0.0	3.1	40.3	0.0	0.0	43.4
TOTAL	**130.6**	**178.3**	**68.0**	**12.4**	**28.5**	**72.5**	**12.0**	**9.0**	**511.3**

- Research on fundamental nanoscale phenomena and processes (PCA 1) remains the largest program component area, with $484 million requested for 2011. Combined with PCA 2 (nanomaterials, $342 million in the 2011 request), this basic research component of the NNI portfolio represents just under half of the total NNI funding request, reflecting the continued importance of sustained Federal funding for fundamental research, feeding the innovation pipeline.
- The Department of Energy now has the largest nanotechnology investment among the NNI agencies, requesting a total of $424 million for 2011. The increase in 2011 over 2010 is due in part to new investments at the Advanced Research Projects Agency-Energy (ARPA-E), as well as increased funding at the Office of Science and the Office of Energy Efficiency and Renewable Energy.
- While funding for PCAs 1 and 2 has been sustained, the fastest-growing PCAs in recent years have been those for environment, health, and safety (PCA 7, increasing from $35 million in 2005 to $117 million in the 2011 request) and nanomanufacturing (PCA 5, increasing from $34 million in 2006 to $101 million in the 2011 request). This is consistent with the NNI's commitment to the goal of supporting responsible development of nanotechnology, and with broad recognition of the importance of both EHS and manufacturing research to realizing the true potential benefits of nanotechnology.

Table. 5. Estimated 2010 Agency Investments by Program Component Area (Doller in Million)

	1. Fundametal Phenomena & Processes	2. Nanometerials	3. Nanoscale Devices & Systems	4. Instrument Research Metrology & Standards	5. Nano-manufacturing	6. Major research Facilities & Instr. Acquisition	7. Environment, Health, and Safety	8. Education & Societal Dimensions	NNI Toal
DOE	103.0	114.5	17.2	21.8	7.0	106.3	2.6	0.5	372.9
NSF	152.6	78.7	43.7	18.3	22.4	37.8	29.8	34.3	417.7
HHS/NIH	48.0	75.9	180.7	18.0	2.2	14.0	17.3	4.6	360.6
DOD	138.8	75.3	148.0	5.9	37.2	28.0	3.1	0.0	436.4
DOC/NIST	22.3	8.4	22.5	19.1	27.2	11.2	3.6	0.0	114.4
EPA	0.2	0.2	0.2	0.0	0.0	0.0	17.1	0.0	17.7
HHS/NIOSH	0.0	0.0	0.0	0.0	0.0	0.0	9.5	0.0	9.5
NASA	0.0	8.6	5.1	0.0	0.0	0.0	0.0	0.0	13.7
HHS/FDA	0.0	0.0	0.0	0.0	0.0	0.0	7.3	0.0	7.3
DHS	0.0	6.5	4.9	0.0	0.3	0.0	0.0	0.0	11.7
USDA/NIFA	1.0	2.0	5.7	0.0	0.2	0.0	1.0	0.5	10.4
USDA/FS	2.0	1.4	0.7	1.1	0.2	0.0	0.0	0.0	5.4
CPSC	0.0	0.0	0.0	0.0	0.0	0.0	0.2	0.0	0.2
DOT/FHWA	0.0	2.0	1.2	0.0	0.0	0.0	0.0	0.0	3.2
DOJ	0.0	0.0	0.0	0.0	0.0	0.0	0.0	0.0	0.0
TOTAL	**467.9**	**373.5**	**429.9**	**84.3**	**96.7**	**197.3**	**91.6**	**39.9**	**1781.1**

Table 6. Proposed 2011 Agency Investments by Program Component Area (Dollars in Million)

	1. Fundametal Phenomena & Processes	2. Nanometerials	3. Nanoscale Devices & Systems	4. Instrument Research Metrology & Standards	5. Nano-manufacturing	6. Major research Facilities & Instr. Acquisition	7. Environment, Health, and Safety	8. Education & Societal Dimensions	NNI Toal
DOE	117.2	121.7	30.4	19.3	20.9	111.3	2.6	0.5	423.9
NSF	140.1	74.3	40.7	16.6	32.2	35.3	33.0	29.0	401.3
HHS/NIH	50.3	80.0	193.8	18.6	2.3	14.4	18.3	4.7	382.4
DOD	151.5	39.3	99.0	2.5	25.1	30.7	0.5	0.0	348.5
DOC/NIST	22.4	8.2	20.2	18.5	20.2	11.2	7.3	0.0	108.0
EPA	0.2	0.1	0.2	0.0	0.0	0.0	19.5	0.0	20.0
HHS/NIOSH	0.0	0.0	0.0	0.0	0.0	0.0	16.5	0.0	16.5
NASA	0.0	8.4	7.4	0.0	0.0	0.0	0.0	0.0	15.8
HHS/FDA	0.0	0.0	0.0	0.0	0.0	0.0	15.0	0.0	15.0
DHS	0.0	6.5	4.9	0.0	0.3	0.0	0.0	0.0	11.7
USDA/NIFA	0.7	1.4	3.8	0.3	0.2	0.0	2.0	0.5	8.9
USDA/FS	2.0	1.4	0.7	1.1	0.2	0.0	0.0	0.0	5.4
CPSC	0.0	0.0	0.0	0.0	0.0	0.0	2.2	0.0	2.2
DOT/FHWA	0.0	1.0	1.0	0.0	0.0	0.0	0.0	0.0	2.0
DOJ	0.0	0.0	0.0	0.0	0.0	0.0	0.0	0.0	0.0
TOTAL	**484.4**	**342.3**	**402.0**	**76.9**	**101.4**	**203.0**	**116.9**	**34.8**	**1761.6**

- There are significant increases requested in nanomanufacturing for 2011—over $26 million, or about 34%, over 2009. A major component of this increase results from the NIST Technology Innovation Program identifying the practical application of advanced materials as an area of critical national need. These advanced materials include nanomaterials, advanced alloys, and composites in manufacturing, resulting in a new 2010 investment in nanomanufacturing.
- NIH's investments in nanotechnology continue to grow, from $40 million in 2001 to $382 million in the 2011 request, as diagnostic and therapeutic applications show increasing promise and maturity.
- Investments in environmental, health, and safety research continue to grow substantially, from $87 million in 2009 (including $12 million in ARRA funds) to a requested $117 million for 2011. The requested EHS investment for 2011 is over triple the figure for 2005 ($35 million). The increases are primarily targeted for those agencies serving as leading agencies for the five R&D categories identified in the NNI *Strategy for Nanotechnology-Related Environmental, Health, and Safety Research* (http.//www.nano.gov/NNI_EHS_Research_Strategy.pdf). Increased funding is provided for some aspects of all five categories of EHS R&D
- Other leading funding categories include the PCAs on nanoscale devices and systems (PCA 3, $402 million in 2011) and major research facilities and instrumentation

acquisition (PCA 6, $203 million in 2011). PCA 3 is growing as nanotechnology R&D moves from passive nanostructures to active devices and complex systems. Sustained investments in the NNI research infrastructure funded under PCA 6, with over 60 centers and user facilities spread across the nation, assure that U.S. researchers and industries have continued access to the world-class facilities and equipment needed to sustain U.S. competitiveness in nanoscale science and technology. Increased funds provided for PCA 6 under ARRA in 2009 were used to upgrade equipment and other facilities at DOE, NSF, and NIST user facilities.

- The largest percentage increases in nanotechnology-related funding for 2011 are at agencies that have had relatively modest nanotechnology investments in the past. Agencies with responsibilities related to EHS issues such as FDA, CPSC, NIOSH, and EPA are requesting significant increases in funding. USDA has also increased its investments in recent years, more than doubling the nanotechnology budget at the National Institute for Food and Agriculture (NIFA, formerly CSREES) since 2007 (from $3.9 million in 2007 to $8.9 million in the 2011 request), and now including a $5.4 million request for nanotechnology R&D at the Forest Service.

Changes in Balance of Investments by Program Component Area (PCA)[5]

P.L. 108-153 calls for this chapter to address changes in the balance of investments by NNI member agencies among the PCAs. These are summarized below for those agencies that are reporting changes for 2010 and 2011.

DOD: The DOD does not foresee major changes in its overall investment portfolio in the nanotechnology program component areas. With the increasing maturation of nanotechnology research, however, increased investment in the broad area of nanomanufacturing (PCA 5) appears likely. Some examples of areas being emphasized include the following: (1) creation of highly uniform arrays of carbon nanotubes with an electronic grade junction to a silicon semiconductor substrate for ultra-lightweight, uncooled infrared detectors, (2) development of bio-enabled fabrication of electronics that bridges top- down and bottom-up approaches to achieve better than 10-nm spatial resolution, and (3) development of electrochemical techniques for patterning metals to produce nanoscale features for antennas, sensors, meta-materials, and catalysts.

DOE: The predominant DOE components of the NNI are research programs and facilities supported by DOE's Office of Science. The investment in 2011 continues to support full operation of the five DOE Nanoscale Science Research Center (NSRC) user facilities (corresponding to PCA 6, major research facilities and instrumentation acquisition) and an extensive array of individual university grants and laboratory research programs. The Energy Frontier Research Centers, larger collaborative efforts in which a portion of the activity relates to nanoscale science, are also continued. In 2010 DOE initiates an Energy Innovation Hub on Fuels from Sunlight, and this support will continue in 2011, with a portion of the activity related to nanoscience. Nanoscience funding increases in 2011 are due partly to small increases across all of these mechanisms, with an emphasis on addressing grand science challenges to achieve atomic and molecular control of matter and energy. However, much of the increase in 2011 over 2010 results from new funding from ARPA-E, the initiation of additional Energy Frontier Research Centers, and the formation of a second Energy Innovation Hub focusing on batteries and energy storage. It is anticipated that a fraction of these activities will be appropriately characterized as nanoscience. The DOE investment in 2011 includes moderate increases across multiple PCAs.

DOT/FHWA: The FHWA's planned increase to its overall nanotechnology investment will be maintained in 2010. 2010 funding will be applied to both nanomaterials (PCA 2) and to nanoscale devices and systems (PCA 3).

CPSC: Formally joining the NNI budget crosscut for the first time in 2011, CPSC is requesting $2 million in increased funding in 2011 in PCA 7 (environment, health, and safety) to allow it to participate with other agencies in researching safety aspects of nanomaterials use in consumer products. Planned 2011 programs include working with other agencies on (1) developing protocols to assess the potential release of airborne nanoparticles from various consumer products and to determine their contributions to human exposure; (2) determining whether nanomaterials can be used for performance improvement in sports safety equipment such as helmets and kneepads without creating other health hazards; (3) expanding consumer product testing using scientifically credible protocols to evaluate the exposure potential from nanosilver in consumer products, with special emphasis on exposures to young children; and (4) working across agencies to assure that shared common public health concerns are met in research studies to determine potential impacts on the public health of nanomaterial use in consumer products.

FDA: Formally joining the National Nanotechnology Initiative budget crosscut for the first time in 2011, FDA will conduct activities in PCA 7 (environment, health, and safety) that support the following agency- wide priorities: (1) laboratory and product testing capacity, (2) scientific staff development and training, and (3) collaborative and interdisciplinary research to address product characterization and safety. Together, these investments will support responsible development of nanotechnology.

NIH: The NIH nanotechnology investment continues to grow (12.5% from 2008 to 2009). Under the American Recovery and Reinvestment Act (ARRA) of 2009 (Public Law 111-5), NIH invested in PCA 7 (environment, health, and safety). In addition to the approximately $13 million that NIH's National Institute of Environmental Health Sciences (NIEHS) invested in nanotechnology EHS research through ARRA (described in Section 4 of this document), NIEHS will provide a $9 million increase to expand its NanoHealth and Safety Initiative. With these 2010 funds, NIEHS will support research to link more precisely the physical and chemical properties of nanomaterials with biological response, thus supplying critical data for hazard and risk assessment. To support the goals of this program, NIEHS is establishing collaborations with the NIH/National Cancer Institute's Nanotechnology Characterization Laboratory for physical characterization of nanomaterials and with the Cancer Biomedical Informatics Grid (CaBIG®) NanoLab for data storage.

NIOSH: NIOSH has focused resources on priorities across the NIOSH Nanotechnology Research Program. The 2011 Budget includes an increase in this program of more than 70 percent over 2010, enabling expansion of research to address major research gaps identified in the NIOSH Nanotechnology Progress Report (http://www.cdc.goviniosh/docs/2010-104/) and in the *Strategic Plan for NIOSH Nanotechnology Research and Guidance* (http.//www.cdc.gov/nosh/docs/2010-105/).

NIST: The President's 2011 Budget for NIST will continue to support research in nanotechnology, including a doubling of the budget for EHS aspects of nanotechnology (PCA 7, from $3.6 million in 2010 to $7.3 million in 2011, including a $4 million new initiative request) and significant new investments in support of the NNI Signature Initiatives focused on nanomanufacturing ($5 million new initiative request) and solar energy. NIST's expanded nanotechnology EHS program will provide the measurement science and technology (reference materials, documentary standards, reference data, instruments, and transferable methods and models) essential for science-based risk assessment and risk management of nanomaterials by regulatory agencies and industry. NIST will target its nanotechnology EHS program to measurements of dynamic physico-chemical and toxicological properties of key nanomaterials and release of these nanomaterials during manufacturing processes and from products throughout full product life cycles. NIST will coordinate and collaborate with other agencies and industry to establish linkages between physico-chemical properties and hazard and exposure effects in biological systems, the environment, and the workplace. NIST's nanomanufacturing research will focus on establishing measurements and methods to efficiently assemble products that integrate together billions or more of nanoscale devices with disparate functions over large areas. The energy-related program will target the nanoscale measurement science and tools needed to accelerate breakthroughs in third-generation photovoltaic solar cell technologies.

NSF: The 2011 NSF request includes an increase to $32.2 million to support new concepts for PCA 5, nanomanufacturing, such as high-rate synthesis and processing of nanostructures, nanostructured catalysts, nanobiotechnology methods, fabrication methods for devices, and assembly of devices into nanosystems and then into larger-scale structures of relevance in industry and in the medical field R&D is aimed at enabling scaled-up, reliable, cost-effective manufacturing of nanoscale materials, structures, devices, and systems. A special focus will be creating active nanostructures and complex nanosystems (PCA 3).

The investment will emphasize (1) new tools for measuring and restructuring matter for production purposes; (2) hierarchical manufacturing of nanosystems by assembling nanoscale components into new architectures and fundamentally new products; (3) manufacturing by design using new computer principles, computer simulations, and nanoinformatics; and (4) hybrid nanomanufacturing, including nanobiotechnology and nanostructured catalysts. An overall goal will be advancing nanomanufacturing methods supporting sustainable development. NSF will strengthen its support for the National Nanomanufacturing Network composed of four Nanoscale Science and Engineering Centers in order to better support innovation and to partner and implement the research results with industry, medical institutions, and other government agencies.

In addition:

- NSF will continue to increase its investment in nanotechnology-related environmental, health, and safety research (PCA 7), by using core programs in relevant areas and a separate program solicitation.
- Partnerships with small businesses in the United States in the areas of nanomanufacturing and commercialization (PCA 5) will be strengthened while maintaining about the same level of NSF investment.
- The NSF 2011 request includes $140.1 million for PCA 1 (fundamental nanoscale phenomena and processes), a reduction of about $12 million compared to 2010. Some funds resulting from this reduction have transitioned into other PCAs as part of the competitive planning process in each NSF directorate.
- Several investments in major research facilities and instrumentation acquisition (PCA 6) will be reduced, since relevant investments made in previous years have taken effect and the research infrastructure is reasonably well established.

USDA/FS: Planned 2011 agency funding for nanotechnology research is focused on cellulosic nanomaterials derived from forest-based materials and remains at the $5.4 million level. The balance of USDA/FS investments by PCA remain unchanged from 2010, with investments in PCAs 1, 2, 3, 4, and 5 (fundamental nanoscale phenomena and processes; nanomaterials; nanoscale devices and systems; instrumentation research, metrology, and standards; and nanomanufacturing, respectively).

USDA/NIFA: The USDA National Institute of Food and Agriculture will continue its balanced investment by supporting research, development, education, and outreach activities in seven of the eight PCAs (see budget tables with PCA distributions). The primary emphases will remain in nanoscale devices and systems (PCA 3) for agricultural and food applications; nanomaterials and nanoscale biomaterials (PCA 2); fundamental nanoscale phenomena and processes (PCA 1); and environment, health, and safety (PCA 7). One new contribution, although small in total dollars, is an increase in PCA 4 for instrumentation research, metrology, and standards for nanotechnology. This is in recognition of the need to characterize engineered nanoscale materials under background environments in various agricultural and food applications.

Utilization of SBIR and STIR Programs to Advance Nanotechnology

As called for by the 21st Century Nanotechnology Research and Development Act, this chapter includes information on use of the Small Business Innovation Research (SBIR) and Small Business Technology Transfer (STTR) programs in support of nanotechnology development. Five NNI agencies—DOD, NSF, NIH, DOE, and NASA—have both SBIR and STTR programs. In addition to these agencies, EPA, NIOSH, NIST, and USDA have SBIR programs. Table 7 shows 2004 through 2008 (the latest year for which data is available) agency funding for SBIR and STTR awards for nanotechnology R&D. USDA/FS reported nanotechnology-related SBIR funding for the first time in 2008.

Table 7. 2004-2008 Agency SBIR and STTR Awards (dollars in millions)

	2004			2005			2006			2007			2008		
	SBIR	**STTR**	**Total**	**SBIR**	**STTR**	**Total**	**SBIR**	**STTR**	**Total**	**SBIR**	**STTR**	**Total**	**SBIR**	**STTR**	**Total**
DOE	6.8	2.8	9.6	7.7	0.4	8.1	18.2	1.6	19.8	17.4	0.8	18.2	13.8	2.7	16.5
NSF	11.9	0.9	12.8	12.1	5.5	17.6	13.9	1.8	15.7	13.4	3.8	17.2	10.5	7.5	18.0
DHHS/NIH	9.3	2.6	11.9	11.1	5.2	16.3	15.1	2.1	17.2	18.4	1.1	19.5	29.3	1.8	31.1
DOD	10.5	6.9	17.4	7.5	5.5	13	12.6	5.6	18.2	8.4	4.2	12.6	19.8	2.3	22.1
DOC/NIST	0.5	0.0	0.5	0.1	0.0	0.1	0.1	0.0	0.1	0.3	0.0	0.3	0.4	0.0	0.4
EPA	0.6	0.0	0.6	1.0	0.0	1.0	1.2	0.0	1.2	0.5	0.0	0.5	0.7	0.0	0.7
DHHS/NIOSH	0.0	0.0	0.0	0.0	0.0	0.0	0.1	0.0	0.1	0.1	0.0	0.1	0.4	0.0	0.4
NASA	7.2	0.6	7.8	6.0	0.0	6.0	12.1	1.5	13.6	11.7	1.5	13.2	6.2	0.8	7.0
USDA/FS	0.0	0.0	0.0	0.0	0.0	0.0	0.0	0.0	0.0	0.0	0.0	0.0	0.3	0.0	0.3
USDA/NIFA	0.8	0.0	0.8	1.0	0.0	1.0	0.7	0.0	0.7	1.1	0.0	1.1	0.6	0.0	0.6
TOTAL	**47.6**	**13.8**	**61.4**	**46.5**	**16.6**	**63.1**	**74.0**	**12.6**	**86.6**	**71.3**	**11.4**	**82.7**	**81.9**	**15.1**	**97.0**

Some NNI agencies (e.g., EPA and NIH) have nanotechnology-specific topics in their SBIR and STTR solicitations. Some (e.g., NIH and NSF) have had topical or applications-oriented solicitations for which many awardees have proposed nanotechnology-based innovations. All told, with this revised data for 2004-2008, the NNI agencies have funded over $390 million in nanotechnology-related SBIR and STTR awards since 2004.

3. Agency Interests in Nanotechnology R&D

Introduction

In 1999 hearings held by the House Subcommittee on Basic Research of the Committee on Science to determine the nation's interest in nanotechnology, Chairman Nick Smith concluded that because "nanotechnology holds promise for breakthroughs in health, manufacturing, agriculture, energy use, and national security... it is sufficient information to aggressively address funding of this field." Shortly afterwards, the White House announced the National Nanotechnology Initiative, and in August 2000, the Subcommittee on Nanoscale Science, Engineering and Technology (NSET) was constituted as part of the NSTC Committee on Technology specifically to facilitate interagency collaboration on nanoscale R&D and to provide a framework for setting Federal R&D budget priorities. The NSET Subcommittee member agencies with budgets dedicated to nanotechnology R&D continue to fund these programs because the work done so far continues to support the early assumptions about the value of this growing scientific endeavor. The agencies describe below their individual interests in nanotechnology R&D, as they collectively contribute by various means to the welfare of the nation and to their respective agency missions and responsibilities.

Consumer Product Safety Commission (CPSC)

As more consumer products employ nanotechnology, concerns are increasing regarding the potential health effects associated with human exposure to this technology. There is a growing use of compounds or materials that have been produced using nanotechnologies that directly manipulate matter at the atomic level and fabrication of materials that could not have been produced in the past. Although these nanomaterials may have the same chemical composition as non-nanomaterials, at the nanoscale these nanomaterials may demonstrate different physical and chemical properties, and they may behave differently in the environment and the human body. Members of the U.S. Congress have stated that they recognize nanotechnology as a new technology utilized in the manufacture of consumer products, and that they expect the Commission to review the utilization and safety of its application in consumer products consistent with the Commission's mission. In support of that mission, CPSC has requested additional funding in 2011 to collect data on nanomaterials use in consumer products

Department of Defense (DOD)

DOD considers nanotechnology to have high and growing potential to contribute to the warfighting capabilities of the nation. Because of the broad and interdisciplinary nature of nanotechnology, DOD views it as an enabling technology area that should receive the highest level of corporate attention and coordination. These priorities are described in the most recent Department of Defense Research and Engineering Strategy, which can be accessed online at http://www.dod.mil/ddre/doc /Strategic_Plan_Final.pdf. DOD Basic Research acknowledges that realizing the potential of nanotechnology is a key research objective for the department. In particular, nanotechnology is an enabling technology for the new classes of sensors (such as novel focal plane arrays), communications, and information processing systems needed for qualitative improvements in persistent surveillance. The DOD also invests in nanotechnology for advanced energetic materials, photocatalytic coatings, active microelectronic devices, and a wide array of other promising technologies. The DOD nanotechnology program is based on coordinated planning and federated execution among the military departments and agencies (components) (e.g., the Defense Advanced Research Projects Agency and the Defense Threat Reduction Agency). Nanotechnology currently represents a scientifically and technologically advanced research theme that has proven and expected value toward enhancing defense capabilities.

DOD does not establish funding targets for nanotechnology, which has proven to be effectively competitive based on its contributions to meeting needs and providing opportunities for future capability. New projects are awarded on a competitive basis, and as a result, the balance of investment can change from historical levels and predictions. Nonetheless, based on the state of nanoscience and nanotechnology research and development, the DOD expects that its primary emphasis will remain in fundamental nanoscale phenomena and processes (PCA 1), nanoscale devices and systems (PCA 3), and nanomaterials (PCA 2). Historically, DOD's Technology Base (Budget Activity 1-3) investment in nanotechnology has been approximately 50% Basic Research, about 40% Applied Research, and about 10% Advanced Technology Development. The bulk of the Small Business Innovation Research (SBIR)/Small Business Technology Transfer (STTR) investment in nanotechnology would be categorized as functionally similar to Applied Research, although with a strong, direct commercializable product or process aspect more analogous to Advanced Technology Development. The current expectation is that these overall percentages will be continued through 2011. As nanotechnology research areas and tools providers continue to mature, growth can be expected in Applied Research and Advanced Technology Development efforts. Maturation of specific products in commercial programs often occurs because of DOD-supported Basic and Applied Research efforts that have increased the knowledge and general technology base. For this reason, continued robust support is important for foundational research in an area that can be valuable to DOD and the nation.

DOD supports competitive research and development in nanotechnology to sustain and increase U.S. technological advantage and general capabilities in warfighting and peacekeeping operations. This provides a very broad front for nanotechnology efforts that are included in DOD programs. An excellent overview of DOD emphasis in nanotechnology may be gained by reviewing the agendas for the upcoming and historical Nanotechnology for Defense Conferences at the website https://www.usasymposium.com/ nano/default.htm. The following foci of the 2009 conference provide a good example of the areas that DOD finds topical: (1) Nanomaterials and Technology for Energy Conversion and Storage, (2) Nanomaterials and Technology for Electronics and EM Sensing, (3) Nanomaterials and Technology for Structural Components, and (4) The Interface between Nano/Bio and its Potential Utility.

Department of Energy (DOE)

DOE views nanoscience and nanotechnology as having a vitally important role to play in solving the energy and climate change challenges faced by the nation. This broad and diverse field of research and development will likely have dramatic impact on future technologies for solar energy collection and conversion, energy storage, alternative fuels, and energy efficiency, to name just a few. DOE has participated in the NNI since its inception in 2001 and maintains a strong commitment to the initiative, which has served as an effective and valuable way of spotlighting needs and targeting resources in this critical emerging area. It continues to provide a crucial locus for interagency communication and collaboration, and an impetus for coordinated planning. Overall investment in physical sciences has been spurred by the NNI, and the research and infrastructure successes have clearly made the United States the world leader in this area, with many other nations seeking to emulate this initiative.

DOE funding spans all eight program component areas of the NNI, with the majority falling into three categories: fundamental phenomena and processes (PCA 1), nanomaterials (PCA 2), and major research facilities and instrumentation acquisition (PCA 6). In the latter category the DOE investment is significantly larger than that of any other agency, due primarily to the planning, construction, and operation of five Nanoscale Science Research Centers (NSRCs) located at DOE laboratories. The NSRCs operate as user facilities, with access based on submission of proposals that are reviewed by independent evaluation boards, and at no cost for non-proprietary work. The NSRCs support synthesis, processing, fabrication, and analysis at the nanoscale and are designed to be state-of-the-art user centers for interdisciplinary nanoscale research, serving as an integral part of DOE's comprehensive nanoscience program that encompasses new science, new tools, and new computing capabilities.

Department of Homeland Security (DHS)

DHS interests in nanoscience and associated areas are mostly materials-related, but the agency is making inroads into device applications as well this year. The applications for the efforts described below are in explosives detection for enhanced security for aviation, mass transit, and first responders:

- *Materials toolbox:* These efforts are focused on the development of materials systems that allow systematic control of chemical and structural features from molecular scales (functional groups) through nano- and microscales. The ability to precisely tune material properties is critical for successful development of improved active sensor surfaces and analyte collection substrates as well as development of novel sensing structures and arrays.

- *Advanced preconcentrators:* The DHS Science and Technology (S&T) Directorate is currently investigating the development of high-performance preconcentrators for use in next-generation detection systems. The focus of these efforts is the development of nano- and microscale materials that enable radio frequency (RF) and optical control of device temperature. To date, several functional prototypes have been demonstrated. Commercialization of these devices is currently being pursued.
- *Advanced sensing platforms:* Work on the development of multimodal carbon nanotube sensing platforms continues with industry partners. The emphasis of these efforts is on development of manufacturing techniques for low-cost sensor platforms.

Department of Justice/National Institute of Justice (DOJ/NIJ)

The NIJ investment in nanotechnology furthers the Department's mission through sponsoring research that provides objective, independent, evidence-based knowledge and tools to meet the challenges of crime and justice, particularly at the state and local levels. New projects are awarded on a competitive basis, and therefore, total investment may change each fiscal year. However, NIJ continues to view nanotechnology as an integral component of its research and development portfolio.

Director of National Intelligence (DNI)

The National Reconnaissance Office (NRO) R&D program focus emphasizes developments in electronics, structural materials, and power generation and energy storage devices:

- In electronics, the program is emphasizing large-scale carbon nanotube (CNT)-based memory, CNTbased logic devices, and CNT field-effect transistors. CNT-based electronics foundry materials and processes are completely compatible with operations in a standard silicon foundry. Additionally, CNT electronics are inherently radiation-hardened, nonvolatile in nature, and extremely low in power consumption because states are retained by van der Waals forces. The 2011 program will emphasize increased density for memory devices, progressing from the current 4 million bit memory to 64 million bit and 512 million bit memories for space-based operations. Such progress represents a 100- fold improvement of current space-survivable memories.
- For structural components, the program will emphasize the use of carbon nanotube threads and yarns to produce conductive wires, impact- and bullet-proof panels, support struts, electromagnetic suppression boxes, and thermal conductive and thermal-electrical converters. Conductive wire research proposes to exceed the conductivity of copper wires while removing up to 80% of the weight of a signal or power harness. Research to produce CNT panels to withstand high-velocity space debris and to stop large-caliber bullets will be conducted. Research to use functionalized CNT yarns to transport heat away from electronic

components and convert the heat to power for space systems will target a 12% energy conversion efficiency.

- Nanotechnology R&D for power generation and energy storage devices emphasizes topics such as quartz nanorods as an improved solar cell cover glass, indium arsenide quantum dot technology for higher-efficiency solar cells, and carbon nanotube electrodes and additives to improve the performance and safety of lithium-ion (Li-ion) batteries. The goal for power generation is to increase solar cell efficiency from the present 33% to 47% for monolithic cells in a space environment. Power storage R&D aims to replace 9 nickel-hydride batteries with a single Li-ion battery capable of 10 years of orbital life.

Environmental Protection Agency (EPA)

The 2011 President's budget request will further EPA's generation of decision-support information to promote the safe development, use, and disposal/recycling of products that contain manufactured nanoscale materials. Sustainability is a theme that runs through the EPA nanotechnology research program. Using green chemistry and life-cycle assessment approaches, EPA is investigating how nanomaterials behave in the environment, how nanomaterial properties may be modified or exposure controls implemented to minimize and manage potential risks from products containing nanomaterials, as well as ways to minimize inputs, including energy usage, during the production of nanomaterials.

EPA's Nanomaterial Research Strategy (http.//www.epa.gov/nanoscience) guides EPA's nanomaterial research. The strategy builds on and is consistent with the scientific needs identified in the 2006 NEHI Working Group report *Environmental, Health, and Safety Research Needs for Engineered Nanoscale Materials* (http.//www.nano.gov/NNI_EHS_research_needs.pdf) and on the 2007 EPA Nanotechnology White Paper (http://www.epa.gov/osa/nanotech.htm). EPA's research activities are coordinated across the U.S. Federal Government through the NNI and internationally within the Organisation for Economic Cooperation and Development (OECD).

Since 2001, EPA has played a leading role in supporting research and setting research directions to develop environmental applications for nanotechnology as well as to understand the potential human health and environmental implications of nanotechnology. The EPA has provided such leadership through its Science to Achieve Results (STAR) grant program and through contracts to small businesses under its Small Business Innovation Research program, both of which have developed a cadre of highly skilled researchers and engineers with expertise in the area of environmental nanotechnology. EPA's own research laboratories are conducting targeted research to inform decisions related to nanotechnology and the environment. EPA collaborates in research with other Federal agencies, as well as internationally by issuing joint grant solicitations with other governments (such as the United Kingdom and the European Commission) and by EPA's own researchers joining other nations in the international testing program being conducted under the auspices of OECD. As the use of nanomaterials increases, EPA will continue to determine how to best leverage advances in nanotechnology to enhance public health and environmental protection, as well as to improve scientific knowledge to understand and avoid any possible negative impacts of nanomaterials on the environment. The goal is to foster technological advances that maximize benefits to society by minimizing environmental impacts.

Food and Drug Administration (FDA)

Nanomaterials often have chemical, physical, or biological properties that are different from those of their larger-scale counterparts. Such differences may include altered magnetic, electrical, or optical properties, structural integrity, and chemical or biological activity. Because of researchers' ability to engineer such properties, nanoscale materials have great potential for use in a vast array of products, including FDA- regulated products intended to protect and promote public health. Also, because of some of their special properties, nanomaterials may pose different issues for toxicologic, safety, and effectiveness assessments. As such, there is a growing need for scientific information and tools to help better predict or detect potential impacts of such changes on human health.

FDA nanotechnology investments in prior and future years are focused on enabling the agency to characterize nanotechnology-based products; develop models for safety and efficacy assessment; and study the behavior of nanomaterials in biological systems and their effects on human health. These investments will support responsible development of nanotechnology.

FDA also continues to foster and develop collaborative relationships with other Federal agencies through the NNI, as well as with sister regulatory agencies, international organizations, healthcare professionals, industry, consumers, and other stakeholders. These collaborations allow information to be exchanged efficiently and serve to identify research needed to address the use nanomaterials in FDA-regulated products. Although FDA activities will likely be relevant to all four NNI goals, FDA efforts are primarily focused on the goal to facilitate responsible development of nanotechnology in three FDA priority areas of (1) building laboratory and product testing capacity, (2) establishing scientific staff development and training, and (3) engaging in collaborative and interdisciplinary research to address product characterization and safety.

National Institutes of Health (NIH)

Nanotechnology is capable of producing tools that can help NIH advance opportunities to develop novel diagnostics and therapeutics as well as tools enabling research at cellular and sub-cellular levels.

The emergence of nanotechnology has opened a new era of design-driven research into the development of unique 3D nanomaterials and nanostructures with the potential for significant clinical impact across a range of diseases and disorders. The NIH continues to expand its support for nanoscale engineering of multifunctional systems for drug and gene therapy, nanostructures for tissue engineering, and a variety of other biomedical applications, in addition to research tools that aid in understanding the underlying causes of diseases. Progress continues in development of sensors that are both selective and highly sensitive, for early diagnosis of disease (when disease is easiest to treat). Some of these sensors involve imaging techniques that pinpoint chemical and biochemical processes that are characteristic of specific diseases; these exams complement current anatomical imaging to provide essential information for early diagnosis. Others may be used as lab tests that identify early markers of diseases such as cancer or heart disease from a small drop of blood or saliva. Multifunctional nanoparticles are being developed to deliver conventional or novel therapeutics directly to the specific tissues or cells in the body that are affected by disease, sparing healthy cells from drug side-effects. Some of these particles may help doctors to detect how much medicine is needed at a particular time, and to dispense only that dose at the time and location in the body where it is needed.

Nanotechnology-based research tools are being used to better understand the causes and course of diseases, and the effects of genetics and environment on individual patients. NIH plays a substantial role in developing understanding of how to design nanoparticles so they can be safe to use both for manufacturing and for medical treatments. NIH research projects are awarded through rigorous peer review by a variety of mechanisms that range from large, multidisciplinary centers as components of agency-designed programs, to projects submitted by individual investigators who compete for funding of their most creative and promising ideas.

Commercialization is facilitated through funding of SBIR/STTR grants and programs at the various NIH institutes that encourage universities and companies to collaborate, and by providing resources and expertise to test novel formulations for safety and biological activity. Ongoing clinical testing of numerous potential products is an encouraging sign that the investment is paying off and continues to be highly productive.

National Institute for Occupational Safety and Health (NIOSH)

Since the inception of the NIOSH nanotechnology program in 2004, NIOSH researchers have conducted pioneering studies on the hazards of manufactured nanomaterials, real-world exposures to workers, and the effectiveness of exposure mitigation techniques such as respiratory protection. The NIOSH Nanotechnology Field Research project (http.//www.cdc.goviniosh/topics/nanotech/field.html), a partnership effort with industry, has been a highly successful and popular resource for hosting organizations and for NIOSH. These studies provide detailed information facilitating risk assessment and management and continue to remind all involved that prudent measures should be taken in mitigating potential exposures to nanomaterials in the workplace.

NIOSH focuses its activities on evaluating the potential hazards of select nanomaterials, assessing human exposures in workplace and potential health risk, evaluating controls and risk management practices for safe handling of nanomaterials, and developing guidance on medical screening and evaluation of workers. Any increase in funding would support research to address the following areas:

- NIOSH toxicology studies have provided better understanding of the ways in which some types of nanoparticles may enter the body and interact with the body's organ systems; the breadth and depth of such research efforts have been limited to a few nanoparticle types.

More types of nanoparticles need to be assessed for characteristics and properties relevant for predicting potential health risks.

- NIOSH field investigators have assessed exposure to engineered nanoparticles in some workplaces, but little data exist on the extent and magnitude of exposures to other types of nanoparticles in workplaces that manufacture or use nanomaterials, nanostructures, and nanodevices.
- NIOSH guidance is a first step toward controlling nanoparticles in the workplace; however, more research is needed on the efficacy and specificity of engineering and work-practice control measures. Significantly more field research is needed to develop guidance based on evaluating possible short- and long-term health risks in nanotechnology workers, and to develop guidance for medical studies and for prospective medical epidemiologic studies.
- The utility of nanotechnology to support development of new technologies (such as sensors, more efficient filters, and better protective materials) that can enhance the protection of workers requires further research and development.

Internationally, NIOSH will expand its support and leadership activities in nanotechnology safety and health in the World Health Organization (WHO), the United Nations Institute for Training and Research (UNITAR), the Organisation for Economic Cooperation and Development (OECD), and the International Organization for Standardization (ISO). In addition, NIOSH is providing support to the International Council on Nanotechnology (ICON) project "Good Nano Guide" (http://goodnanoguide.org/).

National Institute of Standards and Technology (NIST)

Advancing nanoscale measurement science, standards, and nanotechnology is an important component of NIST's mission to promote U.S. innovation and industrial competitiveness. From leading cutting-edge research to coordinating the development of standards that promote trade, NIST's programs in nanotechnology directly impact priorities important to the nation's economy and well-being. The nanotechnology-related research conducted in NIST's laboratories develops measurements, standards, and data crucial to a wide range of industries and Federal agencies, from the development of advanced spectroscopic methods needed to increase efficiency in advanced photovoltaics, to the development of the standard reference materials and data necessary to accurately quantify and measure the presence and impact of nanomaterials in the environment. NIST further supports the U.S. nanotechnology enterprise from discovery to production through its Center for Nanoscale Science and Technology (CNST), established in 2007. The CNST creates the next generation of nanoscale measurement instruments, makes them available through collaboration, and provides rapid access to a world-class, shared-use nanofabrication facility, the NanoFab. NIST's Technology Innovation Program is investing in the development of transformational technologies necessary to advance the large-scale manufacture of nanomaterials. In coordination with the private sector, NIST provides the technical foundation and leadership to the development of international nanotechnology standards—standards that are a key to opening new markets and facilitating trade and commerce.

National Science Foundation (NSF)

The National Science Foundation supports nanoscale science and engineering in all disciplines throughout all its research and education directorates as a means to promote discovery and innovation and integrate various fields of research. NNI funding provides a source of increased interdisciplinarity for about 5,000 active awards that represent more than 10% of the overall NSF portfolio. About 10,000 students and teachers are educated and trained in nanoscale science and engineering each year. Over 100 small businesses were funded to perform research and product development in nanotechnology through the SBIR/STTR programs.

The National Science Foundation supports upstream research and education in all areas of nanoscale science and engineering, leading to a flexible infrastructure and educational pipeline at the national level. It also advances nanotechnology innovation through a variety of translational research programs and by partnering with industry, states, and other agencies. The investment of $418 million in 2010 will support over 4,000 active projects, over 30 large centers grouped in ten networks, and 10,000 students and teachers. About $20 million has been allocated in 2009 for nanotechnology-related SBIR/STTR projects based on merit review; this opportunity will continue in 2011.

Fundamental changes envisioned through nanotechnology have required a long-term R&D vision. NSF sponsored the first initiative dedicated to nanoparticles in 1991, the 1997-1999 program Partnership in Nanotechnology, and it produced the 1999 interagency report *Nanotechnology Research Directions: Vision for Nanotechnology in the Next Decade* that was adopted as an official NSTC document in 2000.

NSF's nanotechnology research is supported primarily through grants to individuals, teams, and centers at U.S. academic institutions. The efforts in team and center projects have been particularly fruitful because nanoscale research and education are inherently interdisciplinary pursuits, often combining elements of chemistry, biology, manufacturing, physics, optics, and photonics, and nearly every other field of basic science. An illustration is the Nanotechnology Computational Network that has reached 100,000 research and education users annually.

A main NSF funding increase in 2011 is for nanomanufacturing to support new concepts for high-rate synthesis and processing of nanostructures, nanostructured catalysts, nanobiotechnology methods, and methods for fabricating devices, assembling them into nanosystems, and then into larger-scale structures of relevance in industry and in the medical field R&D is aimed at enabling scaled-up, reliable, cost-effective manufacturing of nanoscale materials, structures, devices, and systems. The National Nanomanufacturing Network is organized around four centers collaborating with other NNI agencies. Environmental, health, and safety implications of nanotechnology, including development of predictive toxicity of nanomaterials, will be investigated in three dedicated multidisciplinary centers and in over 60 other smaller groups.

U.S. Department of Agriculture, Forest Service (USDA/FS)

Forests are a major strategic asset of the United States and provide critically important economic and environmental benefits, including jobs in rural America, recreation, watershed management, wildlife habitat, and other critically important ecosystem services. Trees, the basis for the U.S. forest products sector, are a sustainable, renewable resource that helps mitigate climate change through the sequestration of carbon—both in the growth of forest biomass and in forest products—and they represent a vast storehouse of renewable feedstock for production of biomass-derived fuels, power, and chemicals. Each year over 200 million tons of wood are converted to over $250 billion of products, employing over 1.4 million Americans and representing about 6 percent of America's manufacturing Gross Domestic Product. Despite the large tonnages used, the United States harvests less than one percent of its total standing forest biomass each year, and the volume of annual growth over harvest ratio in America's forests is over 2:1.

Nanotechnology has enormous promise to bring about fundamental changes and significant benefit to the nation. Nanotechnology offers the way to efficiently and effectively capitalize on a major strategic national asset to make forest-derived materials the "materials of choice for the 21st century." Wood from trees is made up of nanodimensional building blocks that: (1) have strength properties greater than Kevlar* and piezoelectric properties equivalent to quartz, (2) can be manipulated to produce photonic structures, (3) are remarkably uniform in size and shape, (4) possess self-assembly properties, and (5) can be renewably produced in quantities of tens of millions of tons. Recognizing the importance of this emerging opportunity, U.S. Department of Agriculture Forest Service R&D (FS R&D) has taken steps to establish a foundation in forest products nanotechnology research and development. Through participation in the NNI and representation on the NSET Subcommittee, FS R&D is partnering with other Federal entities (e.g., NIST, NSF, DOE, DOD), industry, and academia to develop the precompetitive science and technology critical to the economic and sustainable production and use of new high-value, nanotechnology-enabled, forest-based materials and products.

The U.S. forest products industry, through the American Forest & Paper Association Agenda 2020 Technology Alliance, has signed a memorandum of understanding with the NSET Subcommittee to form a Cooperative Board for Advancing Nanotechnology (CBAN). The U.S. forest products industry is in a unique position to tap the huge potential nanotechnology provides, in two primary ways. First, by becoming a user of nanotechnology materials and components in its products and processes, the industry can upgrade its processes and produce new high-performance consumer products from lignocellulosicbased materials in a safe and sustainable manner. Second, the industry intends to become a producer and developer of novel, sustainable nanomaterials to replace nonsustainable materials such as those from fossil fuels. For example, use of wood-derived nano-dimensional cellulose in nanocomposites will allow the production of much lighter-weight, hyper-strength, multifunctional materials to replace metals and plastics, with widespread application in many industry sectors. Other potential applications include electronic displays and microelectronic devices; clear armor; self-sterilizing and self-healing surfaces; pharmaceutical products; and intelligent wood- and paper-based products with an array of microsensors and nanosensors built in to measure forces, loads, moisture levels, temperature, pressure, chemical emissions, and attack by mold and wood-decaying fungi.

Recognizing that the values of wood and wood-based materials at the nanoscale are virtually untapped, industry technology leaders working with government and academia have developed a vision for nanotechnology and identified three priority focus areas:

- Creating new revenue streams from production of new generations of high-value, high-performance, innovative nanotechnology-enabled products and forest-derived nanomaterials
- Improving the strength/weight performance of traditional paper- and wood-based structural materials
- Developing new value-added, multifunctional features for paper and forest products

FS R&D is continuing to focus on developing its internal wood-based nanotechnology research capacities and to effectively partner with industry, academia, and other Federal entities. The National Nanotechnology Initiative is critical to advancing this public-private-university partnership. In addition, the NNI effectively addresses nanotechnology-focused environmental, health, and safety issues as well as ethical, legal, and societal issues—areas where FS R&D has little or no expertise or capacity to carry out credible work.

U.S. Department of Agriculture, National Institute of Food and Agriculture (USDA/NIFA)

The National Institute of Food and Agriculture (NIFA) of the U.S. Department of Agriculture (USDA) was established by the 2008 Farm Bill to serve the nation's needs by supporting exemplary research, education, and extension that addresses challenges facing the nation. Research enables Americans to develop the knowledge needed to innovatively solve many critical societal issues. Education strengthens schools and universities to train the next generation of scientists, educators, producers, and citizens. Extension brings the knowledge gained and technology development through research and education to the people who need it most—in the United States and around the world.

NIFA's current priority areas include (1) global food security and hunger, (2) climate change, (3) sustainable energy, (4) childhood obesity, and (5) food safety. Nanoscale science and nanotechnology has demonstrated relevance and great potential to enable revolutions in broad agriculture and food systems, including plant production and products; animal health, production, and products; food safety, nutrition, health, and wellness; renewable energy, natural resources, and environment; agriculture systems and technology; and agriculture economics and rural communities. The agency's nanotechnology research has broadly contributed to the NNI PCAs, with primary emphasis on fundamental nanoscale phenomena and processes (PCA 1), nanomaterials (PCA 2), nanoscale devices and systems (PCA 3), and environment, health, and safety (PCA 7).

In particular, NIFA's nanoscale science and nanotechnology research has focused its investment on detection and intervention technologies for enhancing food safety and agricultural biosecurity; effective and safe delivery of bioactives in functional foods for improving human health and wellness; and product traceability, identity preservation, and tracking to embrace the continuous advancement of information technology for better decision making through the Agriculture and Food Research Initiative (AFRI), NIFA's flagship competitive grants program. AFRI also has supported research to assess and analyze the perceptions and acceptance of nanotechnology applications in food and agriculture and nanotechnologybased products by the general public, the agriculture producers, the processing industry, and other stakeholders, using appropriate social science tools. Future opportunities for research investment include using plants or animals (or cells or cultures of plants and animals) to produce specifically constructed nanomaterials at low cost in genetically engineered organisms. NIFA's SBIR program also supports innovative nanotechnology research and development throughout its broad topic areas.

4. Progress towards Achieving NNI Goals and Priorities

NNI Contributions to Administration Goals for the Nation

The NSET Subcommittee's coordination of member agencies' NNI programs is guided by and contributes to many of the Administration's goals for the nation in terms of the economy, energy, the environment, technology, and national defense, among other areas. Details of NNI contributions in these areas are provided in the *NNI Supplement to the President's 2010 Budget;* please see http.//www.nano.gov/NNI_2010_budget_supplement.pdf, pages 13-14.

Activities Relating to the Four NNI Goals

As called for in the 21st Century Nanotechnology Research and Development Act, the 2007 NNI Strategic Plan states that the interagency analysis of progress under the NNI will be provided in the annual NNI Supplement to the President's Budget. As indicated in the plan, the NNI's activities for 2009 and 2010 and plans for 2011 are reported here in terms of how they promote progress toward the four NNI goals. Goal-related activities are in turn reported in terms of two categories of activities: (1) individual agency activities and (2) coordinated activities, including engagement with other agencies and groups and activities external to the NNI, and international activities. A brief report of progress toward the NNI goals in terms of these two categories follows. *The activities described below are only selected highlights of current and planned work of the NNI member agencies and are not an all-inclusive description of ongoing NNI activities.*

Goal 1:Advance a world-class nanotechnology research and development program

The NNI member agencies have sustained a strategic investment in nanotechnology R&D, something that is widely recognized as essential for the development and successful exploitation of any emerging technology. The expanded investment by this Administration in nanotechnology will build on the foundation established over the past ten years and set the stage for transitioning the basic research discoveries into technologies and products for the benefit of our society and the nation's economy.

As indicated in the NNI Strategic Plan, the NNI is (1) sustaining a variety of complementary R&D investment pathways, including single-investigator research, multi-investigator and team efforts, interdisciplinary centers of excellence, and user facilities and networks; (2) mapping the leading edge of R&D by sponsoring topical and strategy-setting workshops; (3) coordinating the NNI research and development investments across multiple agencies; and (4) stimulating collaborations and interactions across agencies, disciplines, industrial sectors, and nations. Examples of efforts toward all four components of this goal are presented below.

Individual Agency Contributions to Goal 1

DOD: Specific examples of DOD investments in Goal 1 are as follows:

- Research in graphene focuses on the study, manipulation, and simulation of charge transport; the engineered synthesis of desired electronic properties; and device fabrication technology that will lead to understanding of the fundamental graphene device physics and technology for advancement of multifunctional electronic devices and establish an enabling science and technology in graphene nanoelectronics for commercial, public, and DOD benefits. The Army Research Laboratory hosted a 2-day workshop on graphene electronics in 2009 as the first in a series of workshops on this topic. The workshop attracted 47 eminent scientists and researchers from academia and several DOD agencies as well as researchers from 5 leading industrial organizations.
- Research is conducted exploiting the ability to integrate unique properties of nanomaterials with conventional, industry-standard microfabrication and nanofabrication technology for the realization of novel actuators for microrobotics, small- and large-scale high-density energy storage and generation systems, and novel micro-initiator concepts for fuse applications. In 2010 this effort is expanded beyond nanoporous silicon to pursue the integration of currently incompatible microelectromechanical systems (MEMS) and nanoenergetic materials and processes, such as ink jet and thin film deposition techniques for an explosive chemical (RDX) and energetic intermetallic compositions.
- Organic nanomaterials for flexible and printed electronics and displays are being pursued by an Army team that has designed and synthesized nanoscale organic molecular pair structures that interact to form a charge separation across the pair, or a charge-transfer complex, that enhances charge transport.
- The area of bandgap engineering involves nanostructures such as quantum wells, superlattices, quantum dots, and nanopillars for a wide variety of applications including infrared imaging, energy generation and harvesting, and laser and light-emitting diode (LED) sources. The first quantum dot solar cells operating in the near-infrared (0.9 micron to 1.1 micron) wavelength range have been demonstrated this past year at the Army Research Laboratory. Material is being provided to the State University of New York at Stony Brook for minority carrier lifetime measurements in type II SLS (strained superlattice) structures for infrared detection.
- At the Army Communications-Electronics Research, Development, and Engineering Center (CERDEC) under the Breakthrough Antenna Technology program, research is being conducted to mature and develop printed conformal antennas utilizing artificial magnetic metamaterials. As part of this objective, a nanomaterials effort will synthesize up to eight different series of magnetic metal nanoparticles with dielectric shell

coatings in kilogram-scale quantities. Included in this effort will be a characterization of each compound with respect to particle distribution and phase uniformity.

- The first significant improvement in thermoelectric power factors and overall thermoelectric performance was demonstrated by Office of Naval Research researchers in a nanocomposite of semi- metallic nanoparticles in a compound semiconductor matrix. The composite exhibited resonant states in the electronic band structure. This material's performance has been independently validated and has attracted industrial interest for automobile waste heat recovery.

DOE: The Department of Energy's support for the National Nanotechnology Initiative continues to come primarily from the DOE's Office of Science through support of fundamental R&D, instrumentation, and construction and operation of major research facilities. In 2010 the 46 new Energy Frontier Research Centers (EFRCs) that DOE awarded and initiated in 2009 are fully operational. These collaborative, multi-investigator, multi-institution research projects bring together interdisciplinary teams from universities, national labs, and industry to focus on energy-relevant fundamental science. A significant fraction of the EFRC research is classified as nanoscience, and that portion is reported in the budget figures provided.

In 2010, DOE will initiate an Energy Innovation Hub focused on solar fuels to continue to encourage collaboration and team science and to connect the research labs to the industrial world. An Energy Innovation Hub focused on batteries and energy storage is requested in 2011. It is expected that some portion of the research activities in both hubs will involve nanotechnology, and that portion has been estimated in the budget figures provided.

NIH (ARRA-Related Activities): NIH has received new funds for 2009 and 2010 as part of the American Recovery & Reinvestment Act (ARRA) of 2009. The ARRA Funds were distributed using various strategies and mechanisms to develop and implement critical research innovations to advance the research enterprise, stimulate future growth and investments, and advance public health and healthcare delivery. NIH initiatives were created to support this endeavor and have resulted in funding of nanotechnology R&D projects. These include the Challenge Grants (RC1), Grand Opportunity (GO) Grants, and Small Business Invcstments (RC3). A complete list of Funding Opportunity Announcements (FOAs) related to the ARRA is available at http://grants.nih.gov/recovery/. For a current listing of ARRA-funded nanotechnology projects, visit the NIH's Research Portfolio Online Reporting Tool (RePORT), accessible via http://report.nih.gov/recovery/ and conduct a keyword search on nanotechnology and ARRA for 2009 within the RePORTer box located at the left column.

Under the Recovery Act, NIH established a new program entitled "Research and Research Infrastructure Grand Opportunities" called the "GO" grants program (RFA-09-004). The purpose of the GO grants program is to support high-impact ideas that lend themselves to short-term funding and may lay the foundation for new fields of investigation. The GO grants program supports large-scale research projects that accelerate critical breakthroughs, early and applied research on cutting-edge technologies, and new approaches to improve the synergy and interactions among multidisciplinary and interdisciplinary research teams. Using this mechanism, for example, the NIH National Institute of General Medical Sciences (NIGMS) made 2-year awards to develop and refine methods for building multifunctional particles and probes that target and deliver chemical agents to locate, activate, image, or medicate targets within a cell or organism. A similar strategy was employed by the National Institute of Biomedical Imaging and Bioengineering (NIBIB), the National Cancer Institute (NCI), and the National Heart, Lung, and Blood Institute (NHLBI) to develop multifunctional delivery systems that combine imaging and therapeutic agents to diagnose and treat diseases (termed "theranostics") through the NIH Challenge Grants in Health and Science Research (RC1) solicitation (RFA-OD-09-003). The purpose of this new program was to fund projects that propose novel research in areas that address specific knowledge gaps, scientific opportunities, new technologies, data generation, or research methods in scientific areas identified by the institutes and centers of NIH through 2010. The NIH National Institute of Dental and Craniofacial Research (NIDCR) also utilized this RC1 mechanism to support R&D in developing new materials such as nanocomposites and smart self-healing materials to enhance restorative properties for dental needs (e.g., adhesives, improved durability, better aesthetics, and maximum biocompatibility) and such as nanostructured surfaces to control endogenous cells and processes *in vivo.* These awards complement NIDCR's recently issued 2009 RFA (Request for Applications) on "Harnessing Inflammation for Reconstruction of Oral and Craniofacial Tissues." Six awards were made at the end of 2009, totaling $3.45 million, to develop advanced bioengineering methodologies and/or nanotechnology-based tools and strategies for resolving acute and chronic inflammation and regenerating such tissues.

NIH/NIEHS increased its investment in understanding the potential health, safety, and environmental effects of engineered nanomaterials by awarding approximately $13 million over a two-year period through ARRA. NIEHS awarded 13 new two-year grants to develop better methods to assess exposure and health effects associated with nanomaterials. Ten of the grants were awarded through the NIH Grand Opportunities program announced in March 2009; see http.//www.niehs.nih.gov/recovery/nanomaterialgo.cfm for details. Three were funded from the NIH Challenge Grants program. All 13 are aimed at developing reliable tools and approaches to determine the impact of engineered nanomaterials on biological systems and health outcomes.

NIH/NCI dedicated $7-8 million of its ARRA funding to pursuing further discovery and clinical translation in cancer nanotechnology. This funding supplemented operation of its Centers of Cancer Nanotechnology Excellence and Nanotechnology Platforms as well as covered selected Challenge (RC1) and Grand Opportunities (RC2) grants.

NIH's National Human Genome Research Institute (NHGRI) selected $1,000 Genome Sequencing Technology Development as its ARRA Signature Project. Nanotechnology is central to several awards made under this and other NHGRI ARRA funding. Included is a study on ethical challenges related to nanotechnology research.

NIH (Non-ARRA Activities): In 2010, NIBIB and NIEHS plan to co-fund a contract proposal in response to a recent solicitation to develop and launch a Federally curated interoperative registry designed to link existing databases that define physical, chemical, and other properties of nanomaterials for the community.

NIBIB also issued a 3-year program announcement in November 2008 entitled "Innovation in Molecular Imaging Probes" (PAR-09-016) that resulted in grants that will engineer nanoprobes or develop nanotechnology-enabled tools to detect and image specific molecular activities *in vivo* for clinical applications. This FOA is a follow-up of a previous NIH Common Fund RFA (RM-04-021 "Innovation in Molecular Imaging Probes").

In 2009, the NIBIB and NHLBI issued a 3-year FOA to support the development of novel, multifunctional delivery systems that can target the delivery of therapeutic agents to specific cells or compartments within cells, *in vivo,* and that possess imaging or sensing capabilities to track delivery and determine therapeutic efficacy. In addition to the real-time assessment of targeted therapy with enhanced efficacy, the identification of quantitative *in vivo* relationships between the carrier and cellular events must be determined, which in turn will lead to new engineering design principles at the nanoscale.

NIST: NIST's research program in nanotechnology is addressing national needs in energy, the environment, healthcare, information technology security, manufacturing, and physical infrastructure, as illustrated by the following examples:

- NIST is broadly applying quantum phenomenon at the nanoscale, including single-photon counting, single and entangled photon sources, and quantum coherence, to develop revolutionary devices for communications, sensing, and spectroscopy.
- Next-generation standards are being developed using nanotechnology, including new electrical resistance standards that exploit the unique quantum properties of graphene, and the first SI-traceable nanoindentation instrument for measuring mechanical properties at the nanoscale.
- NIST has developed a low-cost technique for measuring infrared light that is over 1,000 times more sensitive than existing instruments and is capable of measuring the infrared spectrum at single-photon levels emitted by fragile biomaterials and nanomaterials.
- NIST is developing next-generation chemical measurements with molecular- and atomic-scale resolution and will enhance its capabilities in 2010 with atom-probe tomography.
- NIST's ultralow-temperature (10 mK), ultrahigh-vacuum, high-magnetic-field (15 T), scanning probe microscope laboratory began operation in November 2009, capable of measuring quantum nanoelectronics phenomena with unprecedented spatial and energy resolution.

In 2010 the NIST Technology Innovation Program awarded $22.8 million in grants, primarily to small companies, to advance nanomanufacturing and create nanoscale devices and systems. NIST also awarded multiple ARRA grants supporting over $3 million of university nanotechnology research.

NSF: New efforts planned by NSF in 2011 for Goal 1 include the following:

- Planning and enhancing funding for the programs
    - Nanoelectronics for 2020 and Beyond
    - Nanotechnology for Regenerating the Human Body
    - Nanotechnology for Solar Energy Collection and Conversion
    - Nanostructured Catalysts for Green Manufacturing
    - Nanomanufacturing for Sustainable Development
- Enhancing funding for quantum information systems

USDA/NIFA: NIFA's Agriculture and Food Research Initiative nanotechnology research and development supports the agency's mission and its strategic goals. Currently, AFRI plans to support interdisciplinary, multifunctional projects in five grand societal challenge areas to achieve significant and measurable outcomes. The five goals are: (1) keep American agriculture competitive while ending world hunger, (2) improve nutrition and end child obesity, (3) improve food safety for all Americans, (4) secure America's energy future through renewable biofuels, and (5) mitigate and adapt agriculture to variations in climate. Nanoscale science, engineering, and nanotechnology will provide enabling tools to aid development of innovative solutions to these challenges.

USGS: USGS research in nanotechnology focuses in two areas. Work is underway to examine the fate and effects of nanotubes in sediment-dwelling organisms. Additionally, USGS scientists are examining the fate of nanoparticles in the aquatic environment. Some of this research concerns bacterially mediated production of quantum dots, and the remainder centers on the detection and fate of anthropogenically derived nanoparticles.

Coordinated Activities with other Agencies and other Institutions Contributing to Goal 1

DOD and the Edgewood Chemical Biological Center (ECBC): The Army Aviation and Missile Research Development and Engineering Center (AMRDEC) plans to coordinate with ECBC to test and validate AMRDEC-developed chemical/gas nanosensors.

NIH and the Chinese Academy of Sciences: The National Cancer Institute (NIH/NCI), the National Institute of Environmental Health Sciences (NIH/NIEHS), and the Chinese Academy of Sciences organized the "First Joint U.S.-China Symposium on Nanobiology and Nanomedicine" in Beijing, October 20-22, 2008; the follow-up meeting is to be held in Washington, DC, in 2010. The goal of the symposium was to exchange research experiences and consider possible collaborations in the areas of cancer nanotechnology (prevention, detection, and treatment of the disease), and the effects of exposure to nanomaterials on the general public and the environment. In the United States, over $1.5 billion/year of Federal support is dedicated to nanotechnology research and development (http.//www.nano.gov /NNI_2010_budget_supplement.pdf). The Chinese government has established several nanotechnology research centers, including the National Center for Nanoscience and Technology in Beijing. In response to this rapidly maturing nanotechnology research base in China, this meeting and follow-up conversations concerning possible venues of interactions should produce joint projects leveraging strengths of efforts in both countries and promote exchanges of researchers.

NIH and U.S. and Indian cancer researchers: NCI participated in the 1st Indo-U.S. Cancer Nanotechnology Symposium in New Delhi, February 4-6, 2009. The meeting brought together researchers from India and the United States to discuss possible collaborations, student exchanges, and joint funding opportunities.

NIST, other agencies, industry, and universities: NIST co-sponsors workshops with a wide range of partners to identify measurement needs for nanotechnology. In 2009, workshops were held on Enabling Standards for Nanomaterial Characterization; Frontiers of Characterization and Metrology for Nanoelectronics; Calibrations and Standards for Nanomechanical Measurements; and Nanoscale Measurement Challenges for Energy Applications. In 2010, workshops are planned on Nano-Optics, Plasmonics, and Advanced Materials; and Metrology and Standards for Nanoelectronics.

NRO with DOD (Air Force) and NASA: The National Reconnaissance Office, the United States Air Force Space and Missile Center, and the Air Force Research Laboratory at Wright Patterson Air Force Base are coordinating and conducting joint electronic component R&D. NASA provides early demonstration flights for developed carbon nanotube electronics and radiation testing to qualify electronics for space operations.

NRO with DOD: The Office of the Secretary of Defense, Defense Production Act, monitors R&D success and aids to move pilot-scale plants to full production facilities for carbon nanotube materials for structural components.

NRO with DOE: Power generation and energy storage R&D is conducted with the Department of Energy National Renewable Energy Laboratory.

USDA/FS and industry: The USDA Forest Service has forged a strong working relationship with the U.S. Forest Products Industry, primarily through the American Forest & Paper Association (AF&PA) Agenda 2020 Technology Alliance. AF&PA includes most of the leading forest products companies in the United States. A common research, development, and deployment agenda has been identified with appropriate roles for industry, government, and universities. Industry through the AF&PA Agenda 2020 Technology Alliance is spearheading a drive to raise the visibility of nanotechnology in the forest products sector by (1) having a set of industry priority applications; (2) linking with the U.S. National Nanotechnology Initiative through a Forest Products Industry NNI-NILI-CBAN; (3) developing a core group of Federal laboratories, universities, and industry to advance an industry-led effort that makes use of the expertise that industry has to commercialize nanomaterials and nano-enabled products; and (4) identifying annual funding needs and funding sources.

Goal 2: Foster the transfer of new technologies into products for commercial and public benefit

The NNI member agencies have a number of activities uniquely targeting technology transfer and commercialization, e.g., workshops to gain input from industry and the academic community, SBIR and STTR programs to fund innovations in small businesses, and forefront research infrastructure for use by all nanotechnology researchers, including those from industry. Some positive results from this effort are now evident.

Individual Agency Contributions to Goal 2

DOD: Examples of emphasis and progress in DOD efforts within Goal 2 follow:

- A new detector imaging approach has been demonstrated that converts photons from an incoming optical signal into surface plasmon waves (as sound waves are produced by oscillating air molecules, plasmons are quantized oscillations of the electron gas in a metal). The plasmon waves are formed in a metal film that is typically less than 100 nm thick and bonded to a semiconductor substrate. The plasmon waves are collected by a slit in the metal film, where their energy is focused to produce electron/hole pairs in the semiconductor. The electron/hole pairs, in turn, give rise to a measurable electrical signal. Different polarization signals can be produced from the same imaging chip by fabricating different slit orientations, and multispectral signals can be produced by fabricating an imaging chip that has several different slit sizes. This accomplishment provides a decrease in detector size of at least 50% relative to existing technology at equivalent resolution.
- Munitions enhanced with nano-aluminum powder are being developed to give performance and lethality enhancements in miniaturized munitions needed for the reduced volume of advanced aircraft munitions bays and for weaponization of unmanned air vehicles (UAVs). Munitions using nano- aluminum powder may give improved lethality by improving the airblast and the shrapnel accelerating capability over conventional formulations. Nanoparticles have much higher surface area and are therefore more reactive for a given weight than conventional powders. The reaction rate of nanopowders can be controlled by adjusting the nanoparticle diameter, giving tailorability for optimized airblast and for fragment formation and acceleration. Only a few percent by volume of nano-aluminum powder could previously be added to munitions formulations, which created a barrier to the development, production, and characterization of advanced munitions with significant additions of nano-aluminum powders.

DOE: The Department of Energy supports nanotechnology transfer, industrial manufacturing activities, and commercialization through the Office of Energy Efficiency and Renewable Energy (EERE) and its Industrial Technologies Program (ITP). In 2011, ITP's Nanomanufacturing efforts will be focused on realizing the translation of the tremendous U.S. investment in nanoscience to the industrial sector through large-scale energy-efficient fabrication and functionalization of nanomaterials. This can only be achieved through the accelerated adaptation of advanced nanomanufacturing technologies developed through research, development, and deployment activities among national laboratories, universities, and industry. Examples of nanomaterial fabrication and functionalization technologies within ITP's portfolio include the following:

- Energy-efficient large-scale manufacturing of nanoparticles for photovoltaic and light-emitting diodes (LEDs)
- Nano-composite coatings for improved wear resistances, advanced lithium-ion battery electrodes, and dye-sensitized thin film photovoltaic applications
- Self-assembled, nanostructured carbon for energy storage and water treatment

- Nanocatalysts for diesel engine emission remediation and efficient biofuel production

DOT/FHWA: The Federal Highway Administration's Office of Research, Development, and Technology will maintain an increased investment in 2010 and 2011 through its Exploratory Advanced Research Program. Nanoscale approaches for inhibiting and mitigating corrosion and structural health monitoring via nanoscale sensors and devices are being pursued to enable more durable and efficient infrastructure systems.

NIH: Academic institutions involved in the National Cancer Institute's Alliance for Nanotechnology in Cancer have been very active in forming start-up companies dedicated to the clinical translation and commercialization of cancer diagnostics and therapeutics based on nanotechnology. Overall, 34 companies have been formed since the inception of the program (2005), and 250 patent disclosures have been issued or filed to cover intellectual property of these technologies.

The NIH also reissued the Bioengineering Nanotechnology Initiative for Small Business Innovation Research (PA-09-267) on October 2, 2009. The purpose of this trans-NIH solicitation is to stimulate Small Business Innovation Research that employs nanotechnology to enable the development of diagnostics and interventions for treating diseases. This FOA is designed to bring about fundamental changes to the diagnosis, treatment, and management of an array of diseases and traumatic injuries. To that end, the NIH has identified a wide range of research topics that focus on the engineering of nanoscale structures, processes, and systems as well as other technological innovations (data generation, research methods, etc.) to address biomedical challenges. In parallel, the NIH also reissued the Bioengineering Nanotechnology Initiative for Small Business Technology Transfer grant applications (PA-09-266) for the same purpose.

NIST: NIST's mission to develop measurement technologies in support of technology transfer includes significant nanotechnology investments. New examples include standard reference materials to enable nanoscale mapping of deformation, strain, and stress to ensure performance and mechanical reliability of products manufactured from nanostructured materials; and RM 8820, a new lithographically produced dimensional metrology calibration standard useful for calibrating scanning electron, scanning probe, and optical microscopes.

NRO: NRO-supported R&D for nanotechnology in electronics, structural materials, and power generation and energy storage devices pursues dual-use deliveries to entice commercial interest at the earliest stages. NRO transition goals aim to create commercial "pull" to enable rapid establishment of domestic 21st century manufacturing and to rapidly reduce product cost. CNT electronics developments are monitored by data storage vendors and data center manufacturers; the low-power-consumption and low-heat-production nature of CNT electronics will usher in a new paradigm for space and terrestrial data processing applications.

NSF: Ongoing and planned new activities in support of Goal 2 include the following:

- Strengthen the contribution of translational innovation programs from fundamental research, including in Grant Opportunities for Academic Liaison with Industry (GOALI), Partnerships for Innovation (PFI) , and Industry-University Cooperative Research Centers (IUCRC)
- Partnership with small businesses and large companies in the United States for projects on fundamental research with long-term goals of interest to industry
- Co-sponsor workshops on research directions together with industrial partners
- Increase focus on energy, cyber-physical systems, and synthetic biology research and education

Coordinated Activities with other Agencies and other Institutions Contributing to Goal 2

NNI member agencies and international standards bodies: Technical experts from multiple agencies, including DOD, DOE, EPA, NIOSH, NCI, NIST, and USDA/FS, are working to develop international documentary standards in nanotechnology through the International Organization for Standardization (ISO), as coordinated by the American National Standards Institute's U.S. Technical Advisory Group to ISO TC229 (Nanotechnologies). The Federal agencies' Interagency Committee on Standards Policy (ICSP), chaired by NIST, provides an additional forum for information exchange and coordination on standards policy issues relating to nanotechnology. NNI agencies are also participating in standards development activities within ASTM International's Committee E56 (Nanotechnology), the International Electrotechnical Commission Technical Committee 113 (Nanotechnology Standardization for Electrical and Electronics Products and Systems), and the Institute of Electrical and Electronics Engineers' Nanotechnology Council in developing nanotechnology-related documentary standards. This close cooperation with the American National Standards Institute helps NNI member agencies provide input reflecting U.S. priorities in the international standards arena.

NNI member agencies, OECD: The Department of State chaired the OECD Working Party on Nanotechnology (WPN) to advise on emerging policy issues in science, technology, and innovation related to the responsible development and use of nanotechnology. The WPN is assessing business environments, international research collaboration and coordination, available indicators and statistics, and approaches to public engagement.

NIH and IEEE: The NIH in conjunction with the IEEE Society sponsored a nanomedicine workshop in April 2009 that outlined ongoing research in nanotechnology and industrial applications of nanomedicine.

NIST, VAMAS: NIST led the establishment of the ten-nation Nanoparticle Populations Working Group in the Versailles Project on Advanced Materials and Standards (VAMAS). The group will develop an international consensus in measurement methods as a precursor to documentary standards development, with an initial focus on single-walled carbon nanotube chirality measurements.

NIST, NIH-NCI, FDA: These agencies are collaborating to develop reference materials and relevant measurement protocols for biomedical applications.

NRO, DOE, industry: 2011 will usher in CNT wire testing by Boeing, Tyco Electronics, Northrop Grumman, and Lockheed Martin. Oak Ridge National Laboratory has requested a 12-foot-long, oneinch-diameter cable for high voltage testing; such a cable offers an extraordinarily efficient means to transport power throughout the country.

NRO, A123 Systems: The carbon nanotube battery storage R&D team includes A123 Systems; success in CNT-based batteries offers a potential for 3-to 6-fold improvement in battery technology for electric cars and spacecraft.

NSF and other NNI agencies: The NSET Subcommittee's Nanomanufacturing, Industry Liaison, and Innovation (NILI) Working Group planned and conducted a workshop on Regional, State, and Local Initiatives in Nanotechnology on April 1-3, 2009. A report of the workshop is planned to be published in 2010.

NSF, NIST, and the semiconductor/electronics industry: NSF and NIST are continuing to work jointly with universities and a consortium of companies in the Semiconductor Industry Association (SIA) and the Semiconductor Research Corporation (SRC) on activities to support the Nanoelectronics Research Initiative (NRI) with the goal of demonstrating novel computing devices capable of replacing the complementary metal oxide semiconductor (CMOS) transistor as a logic switch in the 2020 timeframe. R&D activities are taking place at multi-university centers and at NSF nanoscience centers, including the Materials Research Science and Engineering Centers, Network for Computational Nanotechnology hubs, and Nanoscale Science and Engineering Centers (NSECs).

Goal 3: Develop and sustain educational resources, a skilled workforce, and the supporting infrastructure and tools to advance nanotechnology

Significant progress is being made on all three aspects of Goal 3. With respect to education and workforce development, education is among the chief objectives of NNI-funded university research. In addition, specific programs targeted at K-16 education, educating the public about nanotechnology, and improving nanotechnology curricula in U.S. schools and universities have been initiated and are growing in scale and reach. Details are provided in the following text. The extensive network of research centers, user facilities, and other infrastructure for nanotechnology research, which was a key element of the original NNI strategy, is now largely complete.[6]

Individual Agency Contributions to Goal 3

DOD: An instrument is being developed that combines the imaging capability of scanning probe microscopy with the nanomechanical measurement capabilities of instrumented indentation to create a capability to image variations of mechanical properties and topography within a test specimen. This tool will enable researchers to probe the mechanisms that determine responses of nanocomposites to stress at the appropriate length scales, ultimately guiding the design of more efficient structural composites for realistic naval applications.

FDA: In support of Goal 3, FDA plans to increase nanotechnology laboratory testing capacity at its campus in Maryland, and its National Center for Toxicological Research (NCTR)/Office of Regulatory Affairs (ORA) facility in Arkansas, to assess nanotechnology products and their safety. These resources will allow assessment and safety research in support of regulatory decision making. Top-tier priorities will consist of developing:

- Testing methods to assess the safety of products that use nanomaterials (including their stability and interaction with biological systems)
- Testing methods to assess the quality and effectiveness of products that use nanomaterials
- Standards to be incorporated in the preclinical safety assessment of products that contain nanomaterials

In addition, FDA will establish a scientific staff development and training program in nanotechnology. Nanotechnology is a multidisciplinary field requiring expertise in pharmacology, materials science, biology, physics, chemistry, medicine, and toxicology. With the funds requested, FDA will establish a program to train review staff to adequately evaluate the scientific data submitted in support of nanotechnology-containing product regulatory applications. This will include:

- Cooperative training and research work in nanotechnology at national laboratories, government- sponsored research centers, and the Nanotechnology Collaborative Opportunities for Research Excellence (CORE) grants program, described below in the subsection on Goal 4 coordinated activities
- Information exchange and coordination with international regulatory agencies to enable science-based harmonization and sharing of regulatory information and decision processes regarding nanotechnology
- Strengthening FDA nanotechnology scientific training and expertise

NIH: The National Cancer Institute's Alliance for Nanotechnology in Cancer issued administrative supplements to Cancer Nanotechnology Centers of Excellence under the ARRA totaling $4 million. The supplements will be used to maintain and enhance the student and post-doctoral workforce within the centers.

The Nanomedicine Common Fund Initiative includes a strong training component for graduate students and postdoctoral fellows through webinars, postdoctoral exchanges among labs, and trainee sessions at annual meetings. In addition, the NIH Office of the Director provided ARRA funding to support undergraduate and high school students to work in a nanomedicine laboratory and represent the University of California San Francisco at the International Genetically Engineered Machine (iGEM) competition.

In April 2009, NIH hosted NanoWeek, a unique series of events revolving around nanotechnology and its application to medicine and biomedical research. The events varied, with some broadcasted widely to educate the public about basic concepts and their potential applications to diagnosis and treatment of disease. Other events included scientific seminars, poster displays, and in-lab demonstrations of cutting- edge research in nanotechnology and nanomedicine geared to promote information exchange among scientists, clinicians, and engineers working in these disciplines.

NIST: The NanoFab, NIST's national nanofabrication and measurement facility, increased the number of research participants by 30% in the past year and plans to add over $5 million of new tools in 2010.

In 2009, NIST awarded ARRA-funded contracts totaling $6 million for nanomeasurement tools, and it plans to make additional such awards in 2010. Also using ARRA funds, NIST awarded $34.3 million in 2010 to support construction of nanoscience facilities at U.S. universities.

NSF: NSF plans to expand the outreach of the National Center for Nanotechnology Applications and Career Knowledge (NACK, main node at The Pennsylvania State University) to other educational clusters and states in the United States.

Coordinated Activities with other Agencies and Institutions Contributing to Goal 3

DOD: The Aviation and Missile Research Development and Engineering Center is collaborating and leveraging multiple nanotechnology programs with the Oak Ridge National Laboratory (ORNL), Army Research Laboratory (ARL), and the Air Force Research Laboratory (AFRL) to broaden subject matter expertise and utilize specialized laboratories and equipment.

NIH, UC San Diego: The NCI is participating in webinars sponsored by the University of California San Diego. The aim of the webinars is to bring knowledge on medical nanotechnology to academics, physicians, and industrial researchers in the United States and abroad.

NIST, universities, industry: Through ARRA fellowship programs beginning in 2010, NIST will provide opportunities for students, faculty, and industrial researchers to interact with NIST scientists and conduct nanotechnology research in NIST laboratories

NIST, DOE: NIST state-of-the-art X-ray detectors, including an imaging microscope developed through the NIST SBIR program and procured with ARRA funds, have enabled NIST to expand its suite of world-class synchrotron instruments at the National Synchrotron Light Source at Brookhaven National Laboratory. These nanocharacterization instruments, accessible to the scientific community through the DOE User Program, have enabled collaborators from industry to develop advanced materials and devices, including next-generation catalysts.

NRO, national labs, universities, industry: The NRO nanotechnology R&D portfolio is aimed at rapidly developing and establishing a 21st century nanotechnology manufacturing base in the United States. The R&D team includes 22 companies, 2 Government labs, and 19 universities.

NSF, DOE, and DOD: NSF's Nanoscale Informal Science Education (NISE) Network and related projects will continue to provide shared knowledge, resources, and networking for educational activities with DOE and DOD in formal and informal settings.

NSF with DOE (Sandia National Laboratories), NIST, DOD, and NIH: These agencies are coordinating joint student fellowship programs in various areas of nanotechnology.

USDA/NIFA, DOE, and Canada: Following the successful joint grantees' meeting with Canada's Advanced Food and Materials Network (AFMNet) held in conjunction with the DOE Center for Integrated Nanotechnologies (CINT) Nanotechnology User Conference in September 2009 in Santa Fe, New Mexico, USDA/NIFA is planning for a Fall 2010 nanotechnology grantees' meeting with AFMNet and Canada's National Institute for Nanotechnology.

Goal 4: Support responsible development of nanotechnology

The NNI has made significant progress towards the goal of supporting responsible development of nanotechnology. Funding for nanotechnology-related EHS research continues to increase at a rate far in excess of the overall NNI budget growth rate, from $35 million in 2005 to $117 million in the 2011 request. This is only counting the narrowly defined *primary, purpose* EHS R&D. The NNI agencies have reached a strong consensus on a comprehensive strategy to move these investments forward effectively, in line with the roles and responsibilities of the respective agencies involved. The NNI also maintains a strong portfolio of research on ethical, legal, and other societal implications of nanotechnology, along with support for innovative approaches to nanotechnology education at all levels, from K-12 though graduate education and public outreach.

Individual Agency Contributions to Goal 4

DOD: Although no major changes in DOD investment are expected among the four NNI goals, increased focus and coordination on responsible development of nanotechnology will continue to be a growing area of DOD emphasis as applications and manufacturing processes mature. Examples of emphasis and progress in DOD efforts within Goal 4 follow:

- Basic research has created a class of amplifying fluorescent polymers that detect hazardous substances with extremely high sensitivity. In the same research area, a family of nano-engineered, self-segregating additives has been developed that have the ability to transport active materials to the polymer/air interface to self-detoxify contaminated coatings.

- Nanoscale platforms are being developed to enable DNA cloning without the use of cells. A related program uses nanoscale polymerase chain reaction (nano-PCR) with a focus on ultrasensitive, rapid, high-fidelity identification of natural or genetically engineered pathogen strains. These programs will have sensitivities three orders of magnitude better than conventional PCR, permitting single-molecule amplification from complex, for instance, "dirty," environmental samples.

EPA: In support of its Nanomaterial Research Strategy, EPA has initiated a research program to understand which nanomaterials are most likely to enter the environment and how they move and transform within environmental media. This information will help the agency focus its human health and ecological effects research on those nanomaterials and pathways with the most potential for harmful human exposure. The program continues to generate results. For example, STAR researchers at Carnegie Mellon University modified iron oxide nanoparticles to enable them to quickly and easily clean up trichloroethylene, a common groundwater contaminant. Using various cellular models, researchers in EPA's own laboratories have examined the *in vitro* pulmonary toxicity of carbon nanotubes as well as the neurological toxicity of nanoscale titanium dioxide.

FDA: The emerging nature of the science and the potential for rapid development of applications for FDA-regulated products highlights the need for timely development of transparent, consistent, and sound regulatory and scientific evaluation pathways. For 2009, 2010, and 2011, FDA is conducting activities that support the following FDA-wide priorities: (1) build laboratory and product testing capacity, (2) enable scientific staff development and training, and (3) engage in collaborative and interdisciplinary research to address product characterization and safety. Together, these priorities will better enable FDA to support the responsible development of nanotechnology.

NIH/NIEHS: NIEHS has assembled a network of extramural researchers to identify reliable and reproducible assays and methods to assess the biological response to and potential hazard of nanomaterials. Assays under consideration include analysis of cellular functions, immunotoxicity, and genotoxicity. Other projects will examine the utility of proteomics and other data-intensive assays in assessing biological behavior of nanomaterials.

NIOSH: NIOSH received an increase in funding for 2010 in nanotechnology-related environmental, health, and safety research (PCA 7 and Goal 4) as a result of specific language included in the President's Supplemental Budget Request. These funds will provide direct support to an increase in NIOSH's investigations specific to evaluating the potential hazards of select nanomaterials, assessing human exposures in workplaces and associated potential health risks, evaluating controls and risk management practices for safe handling of nanomaterials, and developing guidance on medical screening and evaluation of workers. Specifically, NIOSH will accelerate its activities in the area of risk assessment for carbon nanotubes; develop improved methods to detect and measure select nanomaterials that are being commercialized in high volume, such as nanotubes and metal oxides; evaluate the effectiveness of respiratory protection; and begin to quantitatively evaluate engineering controls for the mitigation of human exposures and environmental releases.

NIST: NIST is developing measurement protocols and analysis methods needed to assess airborne nanoparticle concentrations and exposures. This work will support efforts to measure airborne levels of nanoparticles in research settings for use in safety evaluations as well as in studies of consumer products that may release nanoparticles into the air.

NIST is developing measurement methodologies and associated models for determining dynamic physicochemical and toxicological properties of key nanomaterials in relevant media (e.g., air, water, soil, and biological matrices) and in release of these nanomaterials during manufacturing processes and from products throughout full product life cycles. Dynamic physico-chemical property measurements of nanomaterials will focus on changes in nanomaterial surface attributes (area, composition, morphology, and charge), transformation processes (e.g., aggregation, dissolution, electron transfer), transport, and fate. Toxicological property measurements will focus on genotoxicity and mechanistic cytotoxicity assays, as indicated by biomarkers and other indicators of toxicological response.

In 2010 and 2011 NIST plans to release important reference materials needed for uniform assessment of nanomaterials EHS, including carbon nanotubes and titanium dioxide nanoparticles.

NRO: Nanotechnology R&D in electronics, structural materials, and power generation and energy storage devices are conducted in plants complying with current EPA regulations. Researchers are working with the EPA to help establish improved procedures and rules to protect the health of workers handling nanotechnology substances.

NSF: New efforts by NSF toward Goal 4 are to focus on implications of the next generations of nanotechnology products and productive processes, as well as public participation in nanotechnologyrelated activities, at the NSECs on societal dimensions at Arizona State University and the University of California Santa Barbara.

USDA/NIFA: The NIFA Nanoscale Science and Engineering for Agriculture and Food Program (within AFRI) supports competitive grants to assess and analyze the perceptions and acceptance of nanotechnology applications to food and agriculture and nanotechnology-based products by the general public, agriculture producers and processing industry, and other stakeholders, using appropriate social science tools. This effort started in 2008 and has funded three projects. One of them was awarded to Cornell University, which has produced six "radio blast" pieces for the EarthSky program through interviewing experts about nanotechnology applications and implications for agriculture and food systems. The program is estimated to reach audiences of more than 14 million in the United States and other countries. Another grant funded a research project at the Michigan State University to address public perceptions of emerging applications of nanotechnologies in food and agricultural production, titled "Public Perceptions of Agrifood Nanotechnologies: Using Extension to Assess and Link Stakeholder Knowledge with Public Policies." The project seeks to develop within the Cooperative Extension System the capacity to train extension educators on (1) what nanotechnologies are and what their current and emerging applications are throughout the agrifood system, so that extension agents can speak knowledgeably about these matters with their clientele; and (2) develop in the agents the social science research skills for documenting key public perceptions as revealed through extension agent-client interaction, and for translating/transferring this information to agrifood policy organizations through delivery systems that increase the likelihood of the information being used. The long-term goal of this project is to inform and enhance socially responsive policies for agrifood nanotechnologies at national, state, and population-specific levels.

Coordinated Activities with other Agencies and Institutions Contributing to Goal 4

NNI member agencies, OECD: The DOS has chaired the OECD WPN to advise on emerging policy issues in science, technology, and innovation related to the responsible development and use of nanotechnology. EPA has chaired the OECD Working Party on Manufactured Nanomaterials (WPMN), which is leading an international effort by the 30 OECD member nations and other nonmember nations and organizations to coordinate and collaborate on approaches for better understanding the environmental, health, and safety impacts and the benefits of nanotechnology. In 2008, the WPMN embarked on a cooperative international program to conduct testing on 14 nanomaterials types across 59 environmental endpoints. The EPA is leading the U.S. effort to sponsor the testing of many of these materials.

CPSC, NIOSH, EPA: In 2010, CPSC will establish interagency agreements with NIOSH and EPA to complete a literature search and develop experimental procedures using scientifically credible protocols to quantify releases of and consumer exposures to nanosilver in treated consumer products. Special emphasis will be placed on exposures to young children.

CPSC, EPA, NIEHS, NIST, NNI, NIOSH: The NNI held 3 workshops in calendar year 2009 to review and update the critical research needs that are explicated in the NNI Strategy for Nanotechnology-Related Environmental, Health, and Safety research. In February, NIOSH led a workshop focusing on human and environmental exposure assessment research needs, hosted by CPSC. NIST and EPA led the October workshop on nanomaterials in the environment and related instrumentation, metrology, and analytical methods needs. NIEHS and NIST led the November workshop on research to understand human health implications and the instrumentation, metrology, and analytical methods needed to achieve that understanding. The workshops were planned by representatives of industry, academia, the public health sector, and government. Nearly 200 people participated in each workshop. The results of the workshops will be compiled into reports that will inform the next iteration of the NNI EHS strategy. (For more details on these workshops see http.//www.nano.gov/html/meetings/humanhealth/Series.html.)

DOD: The Army Missile Research, Development, and Engineering Center is currently collaborating with the Missile Defense Agency (MDA) on multiple efforts for weaponry safety and insensitive munitions utilizing nanomaterial-based propellants and developing radiation and temperature nanosensors that offer high performance in extremely harsh environments. AMRDEC is also collaborating with MDA on research for weaponry life extension by developing carbon-fiber composites based on a nanoparticle polymer system in which self-healing and morphing properties restore damaged surfaces to their original condition.

EPA, NSF, USDA/NIFA, EC: EPA's National Center for Environmental Research (NCER), the National Science Foundation, and USDA's NIFA will make awards in 2010 through a joint solicitation entitled, "Increasing Scientific Data on the Fate, Transport, and Behavior of Engineered Nanomaterials in Selected Environmental and Biological Matrices." This solicitation is a collaborative effort with the European Commission. The total estimated U.S. award amount is approximately $4.2 million.

EPA, NSF, NIH/NIEHS, NIOSH, and DOE: Since 2004 EPA's STAR grants program has coordinated interagency requests for applications; agencies involved have included NSF, NIEHS, NIOSH, and DOE. The fourth joint research solicitation by EPA's STAR program was issued in 2007. The solicitation was a collaboration between EPA, NSF, and DOE; over 130 research proposals were received. EPA awarded 15 grants, NSF awarded 6 grants, and DOE awarded 1 grant and made several awards to DOE laboratories. Individual grant awards totaled approximately $9 million. In addition, EPA recently awarded $2 million to study fate and transport in biological systems as part of an NIEHS-led request for applications.

EPA, NSF, NIH/NIEHS, NIOSH, USDA/NIFA, and EC: EPA is leading another interagency solicitation with NSF, NIH/NIEHS, NIOSH, and USDA/NIFA, all in coordination with the European Commission, focused on exposure and safety research data for engineered nanomaterials.

EPA, United Kingdom: EPA's NCER under its STAR research program will also fund two research "e-consortia" teams, each at $2 million for a 4-year period, in collaboration with two research consortia teams from the United Kingdom, on the topics, "Consortium for Manufactured Nanomaterial Bioavailability & Environmental Exposure" and "Transatlantic Initiative for Nanotechnology and the Environment."

FDA, other agencies, and universities: There are many classes of products where the FDA is currently the only agency to conduct research essential for science-based safety and/or risk assessments. While the agency can undertake some self-initiated research, the scale and range of issues and expertise involved necessitate collaborations with academic and government laboratories. Therefore, FDA will establish a CORE program to foster collaborative and interdisciplinary research addressing product characterization and safety. The nanotechnology CORE program will support peer-reviewed research at FDA and in collaboration with academia through grant mechanisms or other approaches. The CORE program will focus on:

- Measurement and detection methods for nanomaterials in FDA-regulated products
- Effects of specific nanomaterial characteristics, such as surface charge, shape, size, and composition, on particle behavior (e.g., distribution in the body) and biological outcomes (e.g., both beneficial effects and toxicities)
- Strategies to better predict, assess, and mitigate potential human health risks

NIH/NIEHS, NIST, NIH/NCI, FDA, NIOSH, Oregon Nanoscience and Microtechnologies Institute, and ASTM International: The NIH/NIEHS in collaboration with NIST held a workshop on Enabling Standards for Nanomaterial Characterization on October 8-9, 2008. The goal was to address the urgent need to accelerate standards development at the pre-standards stage. Sponsors and contributors include the National Cancer Institute, National Institute of Standards and Technology, U.S. Food and Drug Administration, Nanotechnology Characterization Laboratory at NCI-Frederick, National Institute of Occupational Safety and Health, Oregon Nanoscience and Microtechnologies Institute, and ASTM International.

NIH/NCI, NIH/NIEHS, NIH/NIBIB, and NIST: NIH/NCI and NIEHS held a one-day workshop at NIST on October 10, 2008, to establish an International Collaboration for NanoEHS Informatics aimed at developing a federated database system. The Collaboration seeks to establish a global resource of knowledge on nanomaterial characteristics and their biological interactions that can be systematically queried from multiple sites within an interoperable, federated system of databases. Follow-up discussions between NIBIB and NIEHS are underway to define possible venues to address these and related nanobioinformatic needs of the community (NanoHealth Enterprise).

NIH/NIEHS, EPA, and University of Massachusetts: These institutions and other academicians cosponsored the International Conference on the Environmental Implications and Applications of Nanotechnology in June 2009 in Amherst, MA. The conference aim was to provide a valuable forum for scientists, regulators, and policymakers from academia, government, and industry to interact and share new knowledge on the health and environmental impacts of nanotechnology, green nanotechnology, and new environmental applications, and to help direct future research and regulatory needs. The conference discussed global health and safety issues surrounding engineered nanoparticles and nanotechnologies, especially in connection with occupational and environmental health, and provided insights into the latest research results and actions to assure the safety and thereby the future success of nanotechnologies.

NIH, Peoples Republic of China: International coordination of materials characterization and its standardization was a subject of a discussion at the "First Joint U.S.-China Symposium on Nanobiology and Nanomedicine" in Beijing, Peoples Republic of China, October 20-22, 2008 (see p. 31 for additional details).

NIOSH, all other NEHI member agencies: NIOSH led the organizing committee within the Nanotechnology Environmental and Health Implications Working Group to organize the NNI Health and Environmental Exposure Assessment workshop that took place February 24-25, 2009. The workshop aimed to provide an open forum to facilitate effective communication among stakeholders about progress achieved in the human and environmental exposure assessment research category and about the path forward for addressing research needs in this category.

NIOSH, OECD: NIOSH will continue its leadership role in the OECD WPMN steering group on Cooperation on Exposure Measurement and Exposure Mitigation. Within this activity NIOSH is leading the development of globally harmonized protocols for nanomaterial emission assessment, guidance for the use of personal protective equipment, and guidance for the use of engineering controls. NIOSH will also continue conducting experiments and contributing data for carbon nanotubes and nanostructured titanium dioxide (TiO2), silver (Ag), and cerium dioxide (CeO2) to the OECD WPMN sponsorship program for safety testing of nanomaterials.

NIOSH, World Health Organization (WHO): NIOSH will continue developing and disseminating best practices globally for working with nanomaterials in collaborations with occupational safety and health institutions and coordinating activities led by WHO collaborating centers in the nanotechnology area.

NIOSH, International Alliance for NanoEHS Harmonization (IANH): NIOSH will participate in the IANH round-robin testing of toxicological techniques to ensure their reproducibility and to facilitate development of globally harmonized approaches to safety testing.

NIOSH, ICON: NIOSH will continue active participation in the International Council on Nanotechnology's "Nano Good Practices Wild" project, which aims to develop and maintain global best occupational practices for the safe handling of nanomaterials, utilizing a collaborative wild software platform.

NIOSH, CPSC: Under an interagency agreement signed in 2008, NIOSH collaborates with the Consumer Product Safety Commission on studying exposure potential resulting from use of selected spray applications utilizing nanomaterials.

NIOSH, CPSC: In 2010, under an IAG between CPSC and NIOSH, NIOSH will conduct testing to determine the exposure impact of bathroom spray that contains engineered nanomaterials.

NIST, DOD: NIST and the U.S. Army Corps of Engineers co-organized a nanosilver workshop in March 2009 to identify the most pressing measurements and standards needs, including the selection of potential forms of nanosilver for international interlaboratory studies on nanosilver properties and stability.

NIST, CPSC: NIST is leading a coordinated research program with the Consumer Product Safety Commission to determine the release of nanoparticle flame retardants from fabrics and foams.

NIST, FDA, and EPA: These agencies are coordinating the development of benchmark data, measurement methods, and prototype reference materials for nanosilver for biomedical applications, including EHS assessments.

NSF, EPA: To ensure that nanotechnology is developed in a responsible manner, the National Science Foundation and the Environmental Protection Agency continue to fund (over five years, starting in September 2008) two Centers for the Environmental Implications of Nanotechnology (CEIN). The CEINs are an important addition to the NNI and will build on NSF's Center for Biological and Environmental Technologies (CBET) and EPA's STAR grants on nanotechnology. Led by the University of California Los Angeles and Duke University, the CEINs will study how nanomaterials interact with the environment and human health, resulting in better risk assessment and risk mitigation strategies. Each center works as a network, connected to multiple research organizations, industry, and government agencies, and emphasizes interdisciplinary research and education.

External Reviews of the NNI

Public Law 108-153 calls for periodic external reviews of the NNI by the National Nanotechnology Advisory Panel and by the National Research Council (NRC) of the National Academies.

Review by the President's Council of Advisors on Science and Technology, Designated as the National Nanotechnology Advisory Panel (PCAST/NNAP)

The most recent review of the NNI by the PCAST/NNAP was released in April 2008 (available online at httpl/www.whitehouse.gov/sites/default/files/microsites/ostp/PCAST-NNAP-NNI-Assessment-2008.pdf). The following is a list of the recommendations of this chapter, and the NNI responses to each of them:

1. Infrastructure, Management, and Coordination

- 1.1. Ensure continuing support from NNI member agencies and from Congress for NNI multidisciplinary centers, networks and user facilities for nanoscale research. The NNI infrastructure of user facilities, centers, and networks is an unparalleled resource for the nanotechnology R&D community, but it requires sufficient funding to maintain and operate. Having had the foresight to establish these centers, DOE, NSF, NIH, and NIST should provide ongoing strong support for these vital assets. In particular, NSF and NIH should continue to fund large centers and collaborative research groups that enable the multidisciplinary approaches that are essential to advances in basic nanotechnology research. Such multidisciplinary research remains especially vital because many applications will emerge from research at the convergence of historically disparate fields of science and technology. The NNI should continue to foster both "curiosity-driven" researchers and "applications-driven" developers, and their interaction. [pp. 31-32]
    - The funding for PCA 6 on major research facilities and instrumentation acquisition has remained steady over the last seven years with a small increase in 2010. However, contrary to this recommendation, some agencies are not sustaining all of their investments in multidisciplinary research centers. For example, several directorates at NSF are de-emphasizing support for NSF's Nanoscale Science and Engineering Centers, and NSF has discontinued support for its Nanoscale Interdisciplinary Research Team (NIRT) awards.
    - NHLBI and NCI have released solicitations to renew their multidisciplinary centers in 2010 funded through the Programs of Excellence in Nanotechnology

and Cancer Centers of Nanotechnology Excellence programs, respectively. NIH also continues to support large multidisciplinary research groups under the Nanomedicine Common Fund Initiative, and the Bioengineering Research Partnerships, several of which are in the nanotechnology area.

- ARRA funding made available in 2009 has been used in part to upgrade equipment at NSF and DOE user facilities.
- NNI "signature initiatives" proposed for 2011 and beyond (see sidebar at the end of Section 1 of this document) will help improve the balance between "curiosity-driven" and "applications- driven" research funded under the NNI auspices.

- **1.2. Seek to improve infra-agency coordination.** Due to the scope and breadth of nanotechnology's impact, the NNAP recommends that each department and agency with numerous operating divisions impacted by nanotechnology (including DOC, DOD, EPA, HHS, and USDA) establish a cross-cutting task force or some similar mechanism to coordinate and optimize nanotechnology activities and policies more uniformly within the agency as a whole. Where such groups already exist, they should be supported at all levels and should be strengthened horizontally and vertically within the agency. The FDA's Nanotechnology Task Force, which incorporates representation from each of its centers, is a notable example. These intra-agency groups, which should include policy, communications, and budget specialists, will foster improved communication within the agency, across the Federal Government, and with outside stakeholders and agency customers. [p. 32]

    - In addition to FDA, agencies with active internal nanotechnology coordinating groups that have been reported are DOD, EPA, NIH, and NSF.
    - The NNCO and the NSET Subcommittee have also been strongly encouraging agencies to establish nanotechnology portal websites that point to all nanotechnology-related activities within the respective agencies. Agencies with nanotechnology portal websites include DOE, EPA, FDA, NASA, NIBIB, NIOSH, NIST, NSF, OSHA, and USDA/NIFA. The NNI website (http://www.nano.gov; see About the NNI/Departments and Agencies) points to the portal websites that have been established or to the agency's homepage where there is no nanotechnology portal page yet.

- 1.3. Strengthen participation in the NNI by DOC (beyond NISI), DOEd and DOL in light of their respective departmental missions. Interdisciplinary training, broad-based education, workforce preparation, market assessment and evaluation, and standards development are critical challenges for nanotechnology and are essential for the United States to achieve the expected societal and economic benefits of nanotechnology research, development, and commercialization. These needs warrant closer involvement from these agencies in the NNI than has existed to date. [p. 32]

    - Work continues on broadening participation from the agencies PCAST identified, but their participation has already expanded. Within DOC, the NSET Subcommittee and its working groups have had active participants from the Economics and Statistics Administration, the Bureau of Industry and Security (covering export control issues), and the International Trade Administration, in addition to NIST's strong participation. The subcommittee also includes active representation from the Department of Education, which played a strong role in the April 2009 NNI Workshop on Regional, State, and Local Nanotechnology

Initiatives, and which with NSF co-sponsored a workshop on nanotechnology education in 2009. The Department of Education has one active representative, and the Department of Labor has designated two staff members as participants in the NSET Subcommittee.

- 1.4. Coordinate NSET Subcommittee and working groups activities more broadly with related NSTC interagency working groups, especially the Interagency Working Group on Manufacturing R&D, which has identified nanomanufacturing as an area of opportunity. [p. 32]
    - Before this 2008 PCAST report was prepared, the Manufacturing R&D Interagency Working Group co-sponsored a workshop on instrumentation, metrology, and standards for nanomanufacturing, in collaboration with the NSET Subcommittee, NIST, NSF, and ONR. OSTP is currently evaluating ways to better support advanced manufacturing (including nanomanufacturing) and has convened representatives from multiple agencies and interagency working groups (including the NSET Subcommittee) to strengthen coordination and initiate interagency strategy development.
    - The NSET Subcommittee and NNCO have coordinated closely with the Multi-Agency Tissue Engineering Science (MATES) Interagency Working Group (IWG). IWG members assisted in organizing an NNI workshop on nanobiotechnology, for example. During 2009, the NSET Subcommittee and NNCO worked with MATES IWG members in the development of an NNI Signature Initiative proposal on nanotechnology for regenerating the human body.
    - NNCO and the National Coordination Office for Networking and Information Technology R&D (NCO/NITRD) are co-located; staff members confer on matters of mutual interest on a regular basis. There has also been a series of meetings to discuss the intersection between the NNI and NITRD, particularly with respect to quantum information science.
    - The chair of the new NSTC Subcommittee on Quantum Information Science (SQIS) has met with NCO and NNCO staff several times to discuss the intersections between the three groups. The SQIS chair assisted in editing the QIS portion of the proposal for an NNI Signature Initiative on Nanoelectronics for 2020 and Beyond (see the sidebar at the end of Section 1).
- 1.5. Continue to function as the central coordination structure for nanotechnology R&D—including nanotechnology EHS research. The NSET Subcommittee, its working groups, and the NNCO have been, and should continue to be, the locus of coordination for all nanotechnology-related activities. Congress, to the extent that it engages these issues, should support the current interagency coordination and management structure of the NNI through the NSET Subcommittee and the NNCO. [p. 32]
    - The NSET Subcommittee and the NNCO have continued to serve this function, as recommended. For example, as discussed in detail earlier in this document, the NSET Subcommittee has funded, and the NEHI Working Group has organized, a series of NNI workshops in 2009 and 2010 to gather community and stakeholder input on the evolving NNI strategy for nanotechnology-related EHS research. Additional workshops are planned for 2010 to gather input for the

next update of the overall NNI Strategic Plan, due to be completed in December 2010.

2. Standards Development

- **2.1.Participate in the development of voluntary consensus-based standards,** which are crucial to research, commercialization, and safe handling and use of nanotechnology. The NNI agencies, individually and jointly through the NNCO, should participate in and support standards development activities. In particular, the NNI should support U.S. participation at key international standards bodies, such as the International Organization for Standards (ISO). Through Federal agency participation, the NNI should seek to avoid duplication of standards development work at organizations that have overlapping areas of activity. [p. 32]
    - Before this PCAST report was written, the funding agencies of the NSET Subcommittee voted explicitly not to continue to sponsor U.S. participation in ISO TC-229 ("Nanotechnologies") through NNCO financial contributions to ANSI (the only U.S. accredited representative at ISO). NNCO did provide some support for the ANSI TAG to ISO TC-229 in the first two years of its existence, but that support was discontinued at the end of 2007. Instead, the NSET Subcommittee members feel that ANSI should request this support directly from each of the member agencies. ANSI has done so, with mixed results so far. However, the subcommittee is supporting, particularly through travel funding, the extensive activities of the NNCO Director as chair of the ANSI TAG to ISO TC-229.
    - In a similar vein, the NSET Subcommittee provided $125,000 of seed funding to OECD to support the establishment of its Working Party on Nanotechnology (WPN), which is addressing a broad range of economic development issues associated with nanotechnology. EPA has also supported the OECD Working Party on Manufactured Nanomaterials (WPMN), which is addressing EHS issues associated with nanotechnology. The subcommittee's support for the WPN was only a one-time arrangement to help start up the WPN. However, it continues to provide support for NNCO staff to attend WPN meetings.
    - Reports on ISO and OECD nanotechnology-related activities are scheduled as regular agenda items at NEHI and GIN working group meetings.
- **2.2 Develop materials and analytical standards for nanotechnology EHS research.** Such standards are critical to characterizing and monitoring effects of nanomaterials. NIST should lead the development work, in consultation and collaboration with agencies that use such standards, including EPA and FDA. The initial focus should be on nanomaterials that have or are moving toward broad commercial use (e.g., nanoscale gold, silver, metal oxides, carbon nanotubes, and other materials such as those identified in the OECD list of fourteen most common nanomaterials in current applications). [p. 32]
    - The NSET Subcommittee collaborated with NIST in organizing a Workshop on Material Standards for Environmental Health & Safety for Engineered Nanoscale Materials (Sept. 2007, just as this PCAST report was being prepared).

NNCO has been assisting NIST in preparing the report from that workshop for publication.

- As recommended above, NIST has taken the lead in developing standard reference materials for a number of nanomaterials now entering commerce, e.g., gold nanoparticles and carbon nanotubes. NIST has worked closely with NIH/NCI, EPA, and FDA in developing these materials.

- 2.3. Work towards development of minimum data sets of physical and chemical properties of nanomaterials. If properly defined, adoption of a minimum set of data for research on nanomaterials would ensure accurate communication of research results and product properties. It would also enable comparison and reproducibility of EHS testing. This is essential to ensure that evaluations are meaningful and that the assessments of potential EHS impacts are sound. NIST should take a leading role in coordinating efforts to this end among the interagency NNI members. [p. 33]

For clarity, the following response pertains to efforts by the NNI member agencies to work with the ISO Technical Committee 229 on Nanotechnologies, the OECD Working Party on Manufactured Nanomaterials, and the broad EHS research community to come to agreement on a minimum set of parameters/properties of nanomaterials for proper characterization of nanomaterials before use in toxicological research studies. Without such proper characterization, many early studies of the toxicological properties of nanomaterials have led to completely wrong conclusions and inferences.

- This continues to be a critical and largely unmet need. Scientific reviewers and journal editors have been particularly reluctant to require such baseline characterizations as a condition for publication, fostering an avalanche of one-off, non-reproducible studies that offer little clarity on nanomaterial EHS implications nor an adequate basis for regulatory decision making.
- NNI member agencies and the NNCO continue to support work at OECD's WPMN and at ISO Technical Committee on Nanotechnologies' working group on EHS aspects of nanotechnologies to develop such a minimum set of characterization parameters of materials properties for purposes of EHS testing.
- The OECD WPMN is now using a generally agreed upon set of 9 nanomaterials properties for characterizing 14 nanomaterials in relatively common commercial use. ISO TC 229 Working Group 3 also is nearing completion of a Technical Report recommending a similar set of 9 properties for characterization of nanomaterials for toxicological studies.

3. Technology Transfer and Commercialization

- **3.1. Expand efforts to assess national and international innovation and commercialization activities.** While the NNAP commends current and ongoing efforts, the NNI—led by DOC—should identify metrics and obtain data that will allow accurate assessment of the economic impact of nanotechnology development. This will require closer and more coordinated involvement from DOC and continued engagement with OECD to obtain better data at the national and international levels. The downstream impact of nanotechnology development on the economy remains difficult to quantitatively assess, but market projections based on well-defined categories of nanomaterials and devices and specific classes of products will allow for some reasonable estimates, if properly qualified. In any case, since so many industries are involved in nanotechnology development, it appears clear that nanotechnology will have a large economic impact, and continuously monitoring that impact is an important role DOC should play. [p. 33]

- The NSET Subcommittee's support for OECD/WPN discussed above was motivated by the need to expand our understanding of international innovation and commercialization activities. Some preliminary reports from WPN are now becoming available. A representative from the DOC Economic Statistics Administration was initially an active participant in the WPN, and representatives from the DOC International Trade Administration have been engaged in the NSET Global Issues in Nanotechnology working group since 2009.
- NSF has funded a group at the Georgia Institute of Technology to gather information and analyze trends in commercialization of nanotechnology. NSF has also sponsored the National Center for Manufacturing Sciences (NCMS) to gather information on nanotechnology companies in the United States.
- NNCO has just concluded an arrangement with Lux Research to provide NSET Subcommittee members with access to Lux's data on nanotechnology companies and related information on economic impact and commercialization.
- NNCO also funds the Asian Technology Information Program (ATIP) to gather information on nanotechnology R&D and commercialization activities in Asia.
- NSF has just initiated a World Technology Evaluation Center (WTEC) international benchmarking study on nanotechnology. Several other NSET Subcommittee member agencies may also contribute to the study. Preliminary results from the study should be available in late 2010.

- **3.2 Continue to build connections across the innovation ecosystem,** including requiring that multidisciplinary centers partner with industry or with economic development organizations. NSF, NIH, and other major supporters of multidisciplinary nanotechnology-focused centers should explicitly support, maintain, and strengthen cross-sector linkages. [p. 33]

  - As documented elsewhere in this chapter, NNI agencies have strong collaborations with several key industries that view nanotechnology as key to their future, including the semiconductor industry, the chemical industry, and the forest products industry. Efforts to establish similar connections with other industries are continuing.
  - Agencies have required that proposals for interdisciplinary research centers funded under the NNI demonstrate strong industry partnerships. For example, NCI Alliance for Nanotechnology in Cancer centers have been required since their inception in 2005 to collaborate with industry. The centers, based at universities, have been very active in forming start-up companies dedicated to the clinical translation and commercialization of cancer diagnostics and therapeutics based on nanotechnology. Overall, 34 companies were formed since the inception of the program (2005), and 250 patent disclosures were issued or filed to cover intellectual property of these technologies.

- The NSET Subcommittee and several of its member agencies are continuing to sponsor technical workshops covering topics that are of interest across multiple industrial sectors. For example, in May 2008, NIST and USDA/FS sponsored an NNI-affiliated workshop on cross-industry issues in nanomanufacturing.
- The NSET working group on Nanomanufacturing, Industrial Liaison, and Innovation (NILI) continues its work reaching out to regional, state, and locally-based initiatives and to specific industry sectors, including electronics, chemicals, and forest products, through its Consultative Boards for Advancing Nanotechnology.

- **3.3. Educate more scientists and engineers** to become entrepreneurs and skilled technology workers. Transfer of know-how and ideas from university labs to products and processes with commercial value and public benefit occurs primarily through college and university education and research activities. Funding world-class research is the best "training program" for top-notch nanoscale scientists and engineers. The NSF Integrative Graduate Education Research and Traineeship (IGERT) program is a notable model in this regard, particularly with respect to interdisciplinary nanotechnology training and R&D. [p. 33]
    - Earlier sections of this chapter discuss a number of activities aimed at the issues discussed in this recommendation. See Section 4 of this chapter, particularly activities described there relating to Goal 2 (foster the transfer of new technologies into products for commercial and public benefit) and Goal 3 (develop and sustain educational resources, a skilled workforce, and the supporting infrastructure and tools to advance nanotechnology).

4. Environmental, Health, and Safety Implications

- **4.1. Coordinate the nanotechnology EHS strategy with industry and international stakeholders.** EHS research is noncompetitive; therefore, the NNI should coordinate efforts in this area with the activities of industry and other countries so as to avoid duplication and to leverage investments. NNI member agencies should work centrally through the NNCO and/or consensus lead agencies designated in the NNI nanotechnology EHS research strategy to coordinate their respective research activities with other relevant entities. [p. 33]
    - The current (2009/2010) NNI series of EHS workshops are specifically designed to engage stakeholders from all sectors, including industry and international organizations, to gather input on the U.S. strategy for nanotechnology-related EHS research and regulatory activities. The planning teams for each of the workshops conducted in late 2009 and early 2010 include equal participation from industry, non-governmental organizations, universities, and U.S. Government agencies.
    - The OECD/WPMN is aimed at improving international coordination in EHS research and policy development. For example, WPMN is organizing an international collaboration to conduct a series of standard characterization tests on nanomaterials currently in commerce or expected to be in commerce soon.

- ISO TC-229 includes a Working Group 3 on EHS aspects of nanotechnology. This is another appropriate forum for coordinating EHS activities internationally.

- **4.2. Do not segregate implications research and applications research.** In many instances, nanotechnology EHS research cannot be separated from the particular application(s) research andfivm the context for which a specific nanomaterial is intended. Such division is unproductive and neglects the whole benefit of research. Consequently, the NNAP expects that a substantial fraction of nanotechnology research related to EHS will continue to take place under the auspices of agencies that fund applications R&D and may not be uniquely or exclusively identified as nanotechnology EHS research. Risk research that is performed independent of applications development should nevertheless be carried out with consideration of overall risks and benefits associated with the particular material or technology. Furthermore, detailed reporting on the degree of relevance to EHS of such research is not necessarily critical to (and may actually hinder) overall prioritization and coordination. [p. 34]
  - The NNI participating agencies have been criticized (e.g., from the most recent National Academies review of the NNI) for linking implications research to applications research. The December 2008 National Academies Board on Environmental Studies and Toxicology report dismissed a number of the EHS projects reported in the Feb. 2008 NNI EHS strategy document as "not explicitly associated with risk..." (p. 7) and also claimed that EHS research funding figures published by the NNI are overstated because they included projects not directly relevant to nanotoxicology. However, only a small proportion of the funding for NIH projects listed in the 2006 snapshot of NNI EHS projects (included as an appendix to the Feb. 2008 NNI EHS strategy document) was actually counted by the NNI as EHS funding; the remainder was reported in other PCAs appropriate to the health-related applications that those projects are targeting. This is precisely consistent with the 2008 PCAST report recommendation. Virtually every NIH health applications project includes as an integral component some research to make sure that the diagnostic and therapeutic applications being developed are not only effective but are also safe.

- 4.3. Continue developing joint programs among NNI agencies that leverage expertise and resources to conduct nanotechnology EHS research and to support agency missions. The NNI member agencies should proactively seek to collaborate on priority EHS research, where appropriate, in order to expedite progress and take advantage of competency and knowledge that is distributed across the Federal Government. [p. 34]
  - The NNI participating agencies have collaborated on a number of joint interagency efforts to organize and fund EHS workshops listed under the Coordinated Activities section of Goal 4 in this chapter.
  - The NNI participating agencies have funded a series of joint interagency solicitations for nanotechnology-related EHS research in recent years. The agency participation and research foci of these solicitations continue to evolve. Examples include the NSF-EPA Centers for Environmental Implications of Nanotechnology.

- 4.4. Support wide distribution and availability of new nonproprietary information about the properties of nanomaterials. Such information should include methods for risk/benefit analysis that can be implemented by researchers, as well as by developers and manufacturers. [p. 34]

- NIST continues work assessing the physico-chemical properties of key nanomaterials and developing standard reference materials. NCI's Nanotechnology Characterization Laboratory (NCL) also provides this kind of information about nanomaterials being considered for use in clinical applications, with nearly 160 nanoparticles in its assay cascade of physical, *in vitro,* and *in vivo* characterization. (Data and recommended assays by material type are available at http://ncl.cancer.gov.) As indicated earlier, EPA continues its leadership of the international project in OECD's WPMN to develop a wide variety of characterizations of 14 nanomaterials in commercial use.

5. Societal and Ethical Implications

- 5.1. Research on the societal and ethical aspects of nanotechnology should both be integrated with technical ROD and take place in the context of broader societal and ethical scholarship. Societal research should continue to be addressed in conjunction with technical research activities. However, these discussions will also be advanced by involvement of researchers who are primarily engaged in social science, ethics of technology, and other members of the broader academic community with expertise on science, technology, and society. [p. 34]
    - Some centers that NSF has funded to address societal and ethical implications of nanotechnology are co-located with major NNI-funded nanotechnology research centers. Social scientists and nanotechnologists are interacting closely in these centers.
    - The NNCO Director is leading the ISO TC229 Task Group on Consumer and Societal Dimensions of Nanotechnologies. The principal tasks assigned to this group are to explore how all the 32 member countries of the Technical Committee are integrating societal and consumer aspects of nanotechnology into their work on nanotechnology standards and to survey all the other ISO technical committees (some 200) to determine how they are integrating this topic into their standards efforts.

6. Communication and Outreach

- **6.1. Demonstrate more clearly to the public the value of nanotechnology and NNI-supported research and development:** Broader communication and outreach efforts are an essential part of successful innovation. A lack of information and basic understanding of nanotechnology by the general public fosters susceptibility to exaggerated claims and to miscommunications that generate unfounded hopes or fears; these in turn may inhibit future nanotechnology innovation and societal benefit. While communication is a fundamental responsibility of all researchers, a number of specific NNI programs are pursuing efforts to address this both broadly (e.g., the NSF Nanoscale Informal Science Education Network, or NISE Net) and more narrowly, in areas of application (e.g., the model communications efforts of the Alliance for Nanotechnology in Cancer program at NCI). Nonetheless, the

NNI should undertake a more explicit and direct outreach approach to better inform and engage policymakers, stakeholders of all types, and the general public in a dialog as to the application-specific status and associated risk-benefit ratio of relevant near- commercial and commercial nanotechnologies; and to convey the significance of nanotechnology-based capabilities to address grand challenges and future opportunities across industry sectors. Failing to effectively communicate the complete risk-benefit pictures with respect to various specific nanotechnology applications as they exist to date will hinder realization of the significant societal benefits, both demonstrated and promised, of nanotechnology advancements. [pp. 34-35]

- In addition to the various NSF-funded activities mentioned above, NNCO has undertaken a number of activities aimed at communicating better to the public, policymakers, and other stakeholders the potential of nanotechnology. A major redesign of the NNI website, http.//www.nano.gov, is currently underway, including new sections to facilitate public outreach. A brochure on nanotechnology intended for general audiences was completed in 2007 and has been so popular (e.g., with requests for thousands of copies coming from public school systems) that it is now in its third printing—two printings of 10,000 each have been distributed. NNCO is developing a similar brochure discussing the potential of nanotechnology to improve energy conversion, storage, transmission, and efficient utilization. Other brochures on EHS and health applications are planned. The NNCO has a concerted effort underway to develop video footage about various aspects of nanotechnology that will be offered to local TV stations for their use in developing news stories about nanotechnology Finally, NNCO is now completing a first draft of a report documenting commercial spinoffs and research accomplishments arising from the NNI.

- **6.2.Enhance communications efforts within the NNCO.** As an interagency office, the NNCO is well positioned to serve as a central point for much of the communication activity outlined above. In addition, the office also should coordinate among NNI agencies to enhance their agency-specific communication efforts. Member agencies should provide for greater resources to be directed toward these activities. [p. 35]
    - The 2010 NNCO communications plan addresses this recommendation. At the recommendation of the NSET Subcommittee, NNCO staff now includes two full-time communications professionals. NNCO has been reaching out to communications and public affairs offices in the NNI participating agencies but has experienced some challenges in engaging these offices in many of the agencies.

7. Additional EHS Recommendations in the July 2008 PCAST/NNAP Addendum Report

The 2008 PCAST/NNAP report was almost completed when the NSET Subcommittee released its Strategy for Nanotechnology-Related Environmental, Health, and Safety Research in February 2008. Rather than delay the release of the NNAP report, PCAST issued an addendum to its report in July 2008 (http.//www.whitehouse.gov/sites/default/files/microsites/ostp/PCAST-Addendum-Letter.pdf) explicitly reviewing the NNI EHS strategy document. The following is a list of the recommendations from that addendum, and the NNI responses to each of them:

- 7.1. Assess Federal nanotechnology EHS portfolio and update gap analysis against research priorities triennially [p. 4]
  - As part of the process of evaluating and adaptively managing the Federal strategy for nanotechnology-related EHS research, the Office of Management and Budget will issue a new data call to identify and analyze 2009 nanotechnology-related EHS projects funded by NNI participating agencies.
  - As another integral part of this process, the NEHI Working Group, under the sponsorship of the NSET Subcommittee, has launched a series of four workshops treating the five EHS research areas identified in the 2008 Strategy for Nanotechnology-Related EHS Research. After the completion of the fourth workshop in March 2010, NEHI, through its task forces, will use the workshop proceedings to inform and make recommendations to the NSET Subcommittee for adaptively managing the Federal EHS strategy. The first three workshops, held in 2009, assembled experts from academia, state, Federal, and international governments, industry, stakeholders, and the general public to identify the state of the science, review and update the identified EHS research needs, and identify the barriers and gaps to achieving these prioritized research needs.
  - The goal is to release an updated EHS strategy document in late 2010 or early 2011, integrating input from both the updated data call and the workshops described above.
- 7.2. Leverage opportunities to bootstrap identified gap areas and to encourage increased investments elsewhere through collaboration with industry and other countries; encourage broad and ongoing agency participation in such efforts [p. 4]
  - NEHI Working Group member agencies regularly collaborate and coordinate their activities with each other and with other public and private entities, both nationally and internationally, as detailed elsewhere in this chapter. For example, EPA leads the U.S. delegation to the OECD's Working Party on Manufactured Nanomaterials, whose body of work includes testing 14 nanomaterials; DOS has chaired the OECD's Working Party on Nanotechnology; NIST is working with other agencies including EPA, FDA, NIOSH, NCI and DOD to develop international documentary standards in nanotechnology. NSF, EPA, and USDA are coordinating a joint solicitation with the European Commission to promote interagency and international collaboration on EHS research.
- 7. 3. Encourage supported researchers to report on the development of analytical methodologies used in their research so that a series of best practices can evolve for risk assessment and characterization [p . 5]
  - NEHI Working Group member agencies are working together and with other organizations to develop the science and tools to support best practices. The two fall 2009 NNI workshops were deliberately cross-disciplinary efforts in part to identify critical instrumentation, metrology, and analytical methods needed to support best practices in EHS research for the environment and for human health. Presentations included expert comments on the scientific and technical

challenges to creating reliable and reproducible best practices. Other examples include NIH/NCI and NIEHS holding a one-day workshop on nanoEHS informatics to provide a global resource of knowledge on nanomaterial characteristics and their biological interactions; NIH/NIEHS, NIST/NIH/NCI, FDA, NIOSH, Oregon Nanoscience and Microtechnologies Institute, ASTM International, and American National Standards Institute holding a workshop to accelerate standards development; NIOSH participating in the International Alliance for NanoEHS Harmonization round-robin testing of toxicological techniques to ensure their reproducibility and to facilitate the development of globally harmonized approaches to safety testing; and NIST and DOD jointly organizing a workshop on nanosilver to identify the most pressing standards and measurement needs.

- 7. 4. Promote broad and practical use of EHS findings in defining responsible use of nanotechnology in research, manufacturing and commercial application [p. 5]
    - Individual NEHI Working Group member agencies and interagency collaborations are promoting and putting into practice EHS findings. Examples include DOD research to create a class of polymers that detect hazardous substances, EPA's funding of research to use nanoparticles to quickly and easily clean up a groundwater contaminant, a boost in FDA funding to address the increase in numbers and complexity of products involving nanoscale materials under its review, and NSF efforts to focus on the implications for the next generation of nanotechnology and focusing on public participation. The fourth NNI EHS workshop in March 2010 will be on EHS risk management decision making to promote responsible product development while protecting public health. The workshop will also enlarge the EHS discussion to encompass the ethical, legal, and societal implications of nanotechnology.
- 7.5. Increase exposure assessment funding [p. 5]
    - NIOSH is the agency whose research efforts are most closely associated with funding for exposure assessment. In 2009 the agency invested $6.7 million in EHS research, its 2010 budget is estimated to rise to $9.5 million, and it proposes $16.5 million for 2011, a two-and-a-half—fold increase over 2009. These funds will directly support NIOSH's investigations of the potential hazards of select nanomaterials, assessing human exposures in the workplace, evaluating controls and risk management practices for the safe handling of nanomaterials, and developing guidance on medical screening and evaluation of workers.
- 7. 6. Maintain and strengthen agency support and coordination efforts through the NSET subcommittee and NEHI working group [p. 5]
    - As indicated above, agencies are exhibiting strong coordination and collaboration through the NSET Subcommittee and the NEHI Working Group. Examples include the NSET Subcommittee retreat convened in June 2008 (and planned again for June 2010) and the Nanotechnology Signature Initiatives developed by the agency representatives and selected for support in the 2011 budget (including nanotechnology for electronics, solar energy, and sustainable manufacturing).

Review by the National Research Council of the National Academies

The last National Academies review of the NNI, *A Review of the Federal Strategy for NanotechnologyRelated Environmental, Health, and Safety Research* (http://www.nap.edu /catalog.php?record_id=12559), was released in February 2009. The following is a summary of the major findings and recommendations of this chapter, and the NNI responses to each of them:

Findings

- The National Nanotechnology Initiative document Strategy for Nanotechnology-Related Environmental, Health, and Safety Research could be an effective tool for communicating the breadth of federally supported research associated with developing a more comprehensive understanding of the environmental, health, and safety implications of nanotechnology. It is the result of considerable collaboration and coordination among 18 federal agencies and is likely to eliminate unnecessary duplication of their research efforts. [p. 93]
    - NNI representatives appreciate this recognition of some of the merits of the NNI strategy document.
- The Strategy for Nanotechnology-Related Environmental, Health, and Safety Research does not describe a strategy for nano-risk research. It lacks input from a diverse stakeholder group, and it lacks essential elements, such as a vision and a clear set of objectives, a comprehensive assessment of the state of the science, a plan or road map that describes how research progress will be measured, and the estimated resources required to conduct such research. [p. 93]
    - The members of the NEHI Working Group who prepared the February 2008 NNI Strategy for Nanotechnology-Related Environmental, Health, and Safety Research and the members of the NSET Subcommittee who reviewed and approved the document fully recognize the limitations of their first comprehensive Federal strategy for nanotechnology-related EHS research; this strategy explicitly calls for its continuous updating and improvement. However, NNI representatives challenge the factual underpinnings for some of the findings of this National Academies panel, including the bullet above summarizing a number of criticisms of the NNI strategy that appear throughout this National Academies report. The NNCO brought these factual issues to the attention of the National Academies staff after the release of the pre-publication version of this chapter in December 2008. The National Academies addressed some, but not the most substantive, of these issues in its final report, released in February 2009. Subsequently, with review and clearance from the Office of Science and Technology Policy, NNCO published a public rebuttal of some of the National Academies study findings, outlining the factual issues that had not been corrected in the final version of the report, many of which were the basis for major findings and recommendations. For full text of the January 5, 2009, NNI response to the factual misstatements in the NRC review, the reader is referred to the NNCO document on the NNI website, http .//www. nano .gov/Response_ to_NRC_Report_2_13_09.pdf.

Recommendations

- There remains an urgent need for the nation to build on the current research base related to the EHS implications of nanotechnology—including the federally supported research described in the 2008 NNI document—by developing a national strategic plan for nanotechnology- related environmental, health, and safety research. [p. 93; another boldface section on p. 97 outlines what this plan should include]
    - EPA recently funded the Board on Environmental Studies and Toxicology (BEST) at the National Academies to develop a national implementation plan or roadmap for nanotechnologyrelated EHS research that builds on the NNI EHS research strategy. It is expected to engage both public and private stakeholders, identifying roles and responsibilities as well as requisite resources.
- As part of a broader strategic plan, NNI should continue to foster the successful interagency coordination effort that led to its 2008 document with the aim of ensuring that the federal plan is an integral part of the broader national strategic plan for investments in nanotechnology-related environmental, health, and safety research. In doing so, it will need a more robust gap analysis. The federal plan should identify milestones and mechanisms to ascertain progress and identify investment strategies for each agency. Such a federal plan could feed into a national strategic plan but would not itself be a broad, multistakeholder national strategic plan. Development of a national strategic plan should begin immediately and not await further refinement of the current federal strategy. [p. 97]
    - As described previously in the PCAST review sections of this chapter, the NEHI Working Group has organized a series of public workshops in 2009 and 2010 to gather broad research community and stakeholder input on the evolving NNI EHS strategy, as part of the ongoing adaptive management approach laid out in the February 2008 NNI strategy document. One of the main objectives of these workshops is to develop a better analysis of the gaps in the existing Federal EHS investments. As discussed above, EPA has also funded the recommended BEST study to develop a broader national strategic plan.

APPENDIX A. GLOSSARY

Act	Public Law 108-153, the 21st Century Nanotechnology Research and Development Act
AF&PA	American Forest and Paper Association
AFRI	Agriculture and Food Research Initiative (USDA/NIFA)
Agencies	Departments, agencies, and commissions within the Executive Branch of U.S. Federal Government
AMRDEC	Army Aviation and Missile Research Development and Engineering Center (DOD)
ARPA-E	Advanced Research Projects–Energy (DOE)
ARRA	American Recovery and Reinvestment Act
BEST	Board on Environmental Science and Toxicology (National Academies)

BIS	Bureau of Industry and Security (DOC)
CBAN	Cooperative Board for Advancing Nanotechnology
CDC	Centers for Disease Control and Prevention
CEIN	Centers for Environmental Implications of Nanotechnology (EPA and NSF)
CNST	Center for Nanoscale Science and Technology (DOC/NIST)
CNT	Carbon nanotube
CORE	Nanotechnology Collaborative Opportunities for Research Excellence (FDA program)
CPSC	Consumer Product Safety Commission
CSREES	Cooperative State Research, Education, and Extension Service (USDA); it was replaced by the National Institute of Food and Agriculture (NIFA) in 2009
CT	Committee on Technology of the NSTC
DHS	Department of Homeland Security
DHHS	Department of Health and Human Services
DNI	Director of National Intelligence
DOC	Department of Commerce
DOD	Department of Defense
DOE	Department of Energy
DOEd	Department of Education
DOJ	Department of Justice
DOL	Department of Labor
DOS	Department of State
DOT	Department of Transportation
DOTreas	Department of the Treasury
EC	European Commission
EERE	[Office of] Energy Efficiency and Renewable Energy (DOE)
EHS	Environmental, health, and safety
EPA	Environmental Protection Agency
FDA	Food and Drug Administration (DHHS)
FHWA	Federal Highway Administration (DOT)
FOA	Funding Opportunity Announcement
FS	Forest Service (USDA)
GIN	Global Issues in Nanotechnology (NSET Subcommittee working group)
ICON	International Council on Nanotechnology
ISO	International Organization for Standardization
ITC	International Trade Commission
IWG	Interagency working group
NASA	National Aeronautics and Space Administration
NCER	National Center for Environmental Research (EPA)
NCI	National Cancer Institute (DHHS/NIH)
NCL	Nanotechnology Characterization Laboratory (DHHS/NIH/NCI)
NCLT	Center for Learning and Teaching in Nanoscale Science and Engineering (NSF)

NEHI	Nanotechnology Environmental and Health Implications Working Group of the NSET Subcommittee
NHGRI	National Human Genome Research Institute (DHHS/NIH)
NHLBI	National Heart, Lung, and Blood Institute (DHHS/NIH)
NIBIB	National Institute of Biomedical Imaging and Bioengineering (DHHS/NIH)
NIEHS	National Institute of Environmental Health Sciences (DHHS/NIH)
NIFA	National Institute of Food and Agriculture (USDA, replacing CSREES Oct. 1, 2009)
NIH	National Institutes of Health (DHHS)
NILI	Nanotechnology Innovation and Liaison with Industry Working Group of the NSET Subcommittee
NIOSH	National Institute for Occupational Safety and Health (DHHS/Centers for Disease Control and Prevention)
NIST	National Institute of Standards and Technology (DOC)
NNAP	National Nanotechnology Advisory Panel
NNCO	National Nanotechnology Coordination Office
NNI	National Nanotechnology Initiative
NNIN	National Nanotechnology Infrastructure Network (NSF program)
NPEC	Nanotechnology Public Engagement and Communication Working Group of the NSET Subcommittee
NRC	Nuclear Regulatory Commission (also National Research Council of the National Academies)
NRO	National Reconnaissance Office
NSEC	Nanoscale Science and Engineering Centers (NSF program)
NSET	Nanoscale Science, Engineering, and Technology Subcommittee of the NSTC
NSF	National Science Foundation
NSRC	Nanoscale Science Research Centers (DOE program)
NSTC	National Science and Technology Council
NTP	National Toxicology Program (DHHS)
OECD	Organisation for Economic Co-operation and Development
OMB	Office of Management and Budget (Executive Office of the President)
OSTP	Office of Science and Technology Policy (Executive Office of the President)
PCA	Program Component Area
PCAST	President's Council of Advisors on Science and Technology
RFA	Request for Applications
SBIR	Small Business Innovation Research Program
SI	System Internationale (International System of Units)
STAR	Science to Achieve Results (EPA)
STTR	Small Business Technology Transfer Research Program
USGS	U.S. Geological Survey (Department of the Interior)
USPTO	U.S. Patent and Trademark Office (DOC)
USDA	U.S. Department of Agriculture

WHO	World Health Organization
WPMN	Working Party on Manufactured Nanomaterials (under the Chemicals Committee of the OECD)
WPN	Working Party on Nanotechnology (OECD)

End Notes

[1] General note: In conformance with Office of Management and Budget style, references to years in this chapter are to fiscal years unless otherwise noted.

[2] http.//www.nano.gov/NNI_Strategic_Plan_2007.pdf. Note that this plan will be updated in 2010.

[3] Section 7511 of the Food, Conservation, and Energy Act of 2008 (FCEA) established within the Department of Agriculture the National Institute of Food and Agriculture (NIFA) and transferred all authorities of the Cooperative State Research, Education, and Extension Service (CSREES) to NIFA not later than October 1, 2009.

[4] *Strategy for Nanotechnology-Related Environmental Health, and Safety Research* (February 2008): http.//www.nano.gov/NNI_EHS_Research_Strategy.pdf. NNI EHS research is defined as research whose primary purpose is to understand and address potential risks to health and the environment that engineered nanomaterials may pose.

[5] Changes are as compared to NNI investments described in the NNI Supplement to the President's 2010 Budget, http.//www.nano.gov/html/res/pubs.html.

[6] See a detailed discussion of the U.S. NNI infrastructure in the *NNI Supplement to the President's FY 2008 Budget,* http.//www.nano.gov/NNI_O8Budget.pdf, including a map that shows the location of all the centers, networks, and user facilities (p. 21) and a list of participating academic institutions and national laboratories (pp. 29-33).

In: Nanotechnology: Looking Ahead
Editor: Julian S. Moser

ISBN: 978-1-61761-785-0

Chapter 4

REPORT TO THE PRESIDENT AND CONGRESS ON THE THIRD ASSESSMENT OF THE NATIONAL NANOTECHNOLOGY INITIATIVE

Executive Office of the President

ABOUT THE PRESIDENT'S COUNCIL OF ADVISORS ON SCIENCE AND TECHNOLOGY

The President's Council of Advisors on Science and Technology (PCAST) is an advisory group of the Nation's leading scientists and engineers, appointed by the President to augment the science and technology advice available to him from inside the White House and from cabinet departments and other Federal agencies. PCAST is consulted about and often makes policy recommendations concerning the full range of issues where understandings from the domains of science, technology, and innovation bear potentially on the policy choices before the President. PCAST is administered by the White House Office of Science and Technology Policy (OSTP).

For more information about PCAST, see www.whitehouse.gov/ostp/pcast

Co-Chairs

John P. Holdren
Assistant to the President for
Science and Technology
Director, Office of Science and
Technology Policy

Eric Lander
President and Director
Broad Institute of Harvard and MIT

Harold Varmus
President
Memorial Sloan-Kettering
Cancer Center

Members

Rosina Bierbaum
Dean, School of Natural Resources
and Environment
University of Michigan

Christine Cassel
President and CEO
American Board of Internal Medicine

Christopher Chyba
Professor, Astrophysical Sciences and
International Affairs
Director, Program on Science and Global
Security Princeton University

S. James Gates, Jr.
John S. Toll Professor of Physics
Director, Center for String and Particle Theory
University of Maryland, College Park

Shirley Ann Jackson
President
Rensselaer Polytechnic Institute

Richard C. Levin
President
Yale University

Chad Mirkin
Rathmann Professor, Chemistry, Materials
Science and Engineering, Chemical and
Biological Engineering and Medicine
Director, International Institute of

Nanotechnology
Northwestern University

Mario Molina
Professor, Chemistry and Biochemistry
University of California, San Diego
Professor, Center for Atmospheric Sciences at
the Scripps Institution of Oceanography
Director, Mario Molina Center for Energy and
Environment in Mexico City

Ernest J. Moniz
Cecil and Ida Green Professor of Physics and
Engineering Systems
Director, MIT's Energy Initiative
Massachusetts Institute of Technology

Craig Mundie
Chief Research and Strategy Officer
Microsoft Corporation

Ed Penhoet
Director, Alta Partners
Chairman of the Board, Immune Design
Chairman of the Board, Metabolex

William Press
Raymer Professor in Computer Science and
Integrative Biology
University of Texas at Austin

Maxine Savitz
Vice President
National Academy of Engineering

Barbara Schaal
Chilton Professor of Biology
Washington University, St. Louis
Vice President, National Academy of Sciences

Eric Schmidt
Chairman and CEO
Google, Inc.

Daniel Schrag
Sturgis Hooper Professor of Geology

Professor, Environmental Science and Engineering
Director, Harvard University-wide Center
for Environment
Harvard University

David E. Shaw
Chief Scientist, D.E. Shaw Research
Senior Research Fellow, Center for
Computational Biology and Bioinformatics,
Columbia University

Ahmed Zewail
Linus Pauling Professor of Chemistry and Physics
Director, Physical Biology Center
Professor, Chemistry and Physics
California Institute of Technology

Staff

Deborah Stine
Executive Director, PCAST

Mary Maxon
Deputy Executive Director, PCAST

Executive Report

The National Nanotechnology Initiative 2010 Third Assessment along with Recommendations of the National Nanotechnology Advisory Panel

The National Nanotechnology Initiative (NNI) is the U.S. Government's crosscutting program that coordinates Federal research and development (R&D) activities in nanoscale science, engineering, technology, and related efforts among various participating agencies. The Federal Government launched the NNI in FY 2001 with an initial $500 million budget to accelerate the development of nanotechnology. Over the ensuing 10 years, with cumulative Federal spending of $12 billion, the NNI has played a key role in positioning the United States as the world leader in both nanotechnology R&D and commercialization. The NNI has also catalyzed State activities that leverage Federal investments with a focus on economic growth and job creation. Indeed, nanotechnology appears slated to become an important contributor to the economic growth of the United States over the coming decade and beyond.

Today, the NNI is a multi-agency initiative that now includes 25 Federal agencies, 15 of which in FY 2011 will have their own individual budgets for nanotechnology R&D. The NNI is managed within the framework of the National Science and Technology Council (NSTC), the Cabinet-level council by which the President coordinates science and technology policy across the Federal Government. The Nanoscale Science, Engineering, and Technology (NSET) Subcommittee of the NSTC coordinates the planning, budgeting, program implementation, and review of the NNI. The National Nanotechnology Coordination Office (NNCO) provides technical and administrative support to the NSET Subcommittee, serves as a central point of contact for Federal nanotechnology R&D activities, and engages in public outreach on behalf of the NNI. The NNCO also serves as a liaison to academia, industry, professional societies, foreign organizations, and others to exchange technical and programmatic information. Additionally, the NNCO coordinates preparation and publication of NNI interagency planning, budget, and assessment documents, and maintains the NNI Web site, www.nano.gov.

The 21st Century Nanotechnology Research and Development Act of 2003 (Public Law 108-153) calls for a National Nanotechnology Advisory Panel (NNAP) to periodically review the Federal nanotechnology R&D program, that is, to review the NN I. The President's Council of Advisors on Science and Technology is designated by Executive Order to serve as the NNAP. PCAST's first review of the NNI was issued in 2005, the second in 2008.

In this chapter, PCAST, serving in its role as the NNAP, assessed the effectiveness of the NNI over the past two years and since its inception. PCAST's observations, conclusions, and recommendations presented here are based on the analysis of its 2010 NNI Working Group, consisting of 3 PCAST members and 12 additional nongovernmental experts in nanotechnology. The Working Group's deliberations were informed by discussions with government officials, industry leaders, and technical experts from the wide range of fields affected by nanotechnology. Before beginning its review, the Working Group co- chairs received input from the other members of PCAST, relevant Congressional committee staff, and staff from the Government Accountability Office (GAO), Office of Management and Budget (OMB), the National Research Council (NRC), and OSTP. Based on that input, the Working Group decided that there were three overarching categories that would prove most useful for assessing NNI performance and for arriving at valuable and actionable recommendations to ensure that NNI can succeed in the many roles it has to play. Those categories are:

1. **Program Management**—An appraisal of how well NNI leadership has performed with respect to the roles it has been tasked to carry out.
2. **Nanotechnology Outcomes**—An analysis of what the Federal nanotechnology investment has delivered and recommendations to enhance the outcomes, especially economic outcomes.
3. **Environment, Health, and Safety (EHS)**—An assessment of NNI's performance in helping to orchestrate the identification and management of potential risks associated with nanotechnology, with particular attention paid to reviewing progress the N NI has made in following through on recommendations made in the 2008 NNAP review of the NNI.

Because 2010 marks the tenth anniversary of the NNI, the panel decided it would be appropriate to conclude with a forward-looking chapter that discusses how nanotechnology might contribute to important societal needs and goals in the coming years.

Program Management

In its review of program management, the NNAP determined that the NNI has distinguished itself during its first decade as a successful cooperative venture, now involving the participation of 25 Federal agencies. The NNAP believes that the NNI has been well organized and managed, but that there are several steps that OSTP can take to enable the NNI over the next decade to fully exploit the opportunities offered by the development of nanoscience and nanotechnology, particularly in regard to commercialization efforts.

More specifically, the N NAP believes that the NN I would benefit by OSTP designating more resources for the NNCO to coordinate activities across the NN I and by authorizing dedicated funding of approximately $5 million annually (0.3 percent of agency contributions to NNI funding) to support that mission. A strengthened and suitably funded NNCO should be the coordinating agency to ensure several desirable improvements in managing the NNI, including:

- Enhancing communications with the business community;
- Facilitating technology transfer;
- Providing information on available nanotechnology resources to both the public and private sectors. The NNI has created outstanding resources for the nanotechnology R&D community through its investments in infrastructure such as shared user facilities, research centers and networks, and education centers and networks. Accessible information on these facilities and coordination of their use would ensure optimal leverage of these considerable investments;
- Collaborating with stakeholders on enabling programs such as metrology; standards including size, shape and composition of nanomaterials, and databases of physical and chemical properties of nanomaterials; and manufacturing safety; and
- Engaging in closer and more frequent interactions with state initiatives, which could provide important leverage of resources for the NNI.

In addition, the NNAP believes that as programs are developed within the NNI, the NNCO should track relevant metrics to measure the outcomes and impacts of NNI programs. To assess systematically the outcomes of NNI investment, it would be desirable, for example, to be able to measure the value of all products with nanotechnology components, as well as the value of those components. Following both of these measures over time would give a reasonable picture of the importance of nanotechnology in the economy. Such measurements would in turn enable an estimate of the number of jobs created by nanotechnology, as well as the social rate of return to NNI investment. Rather than relying on funding agencies for such estimates, it would be most appropriate to lodge responsibility with a statistical agency such as the Bureau of Economic Analysis in the Department of Commerce.

A strengthened NNCO will be better positioned to monitor metrics that assess the NNI's impact, such as on job creation, commercial activity, private sector investment including venture funding, advanced degrees, international competitiveness, and the development of methodology for assessing plausible risks associated with nanotechnology.

PROGRAM MANAGEMENT RECOMMENDATIONS

Strengthen the NNCO

The NNCO should broaden its impact and efficacy and improve its ability to coordinate and develop NNI programs and policies related to those programs. OSTP should facilitate these improvements by taking the following actions:

- Require each agency in the NNI to have senior representatives with decision-making authority participate in coordination activities of the NNI.
- Strengthen the NNCO to enhance its ability to act as the coordinating entity for the NNI.
- Mandate that the NNCO develop metrics for program outputs and that it works with the Bureau of Economic Analysis to develop metrics and to collect data on the economic impacts of the NNI.
- Appoint two individuals to the NNCO to lead interagency coordination of efforts in the areas of EHS research and standards development, respectively.
- Dedicate 0.3 percent of NNI funding to the NNCO to ensure the appropriate staffing and budget to effectively develop, monitor and assess NNI programs.

Focus on Commercialization

In a budget planning process coordinated by OSTP, each agency should continually re-evaluate its NNI balance of investments among the Program Component Areas (PCAs), with an enhanced focus on commercialization, which would include maintaining the current level of investment in research and doubling the investment in nanomanufacturing (PCA5) over the next five years.

Signature Initiatives

Each Signature Initiative's lead agency should develop coordinated milestones, promote strong educational components, and create public-private partnerships to leverage the outcomes of the Initiatives. Each lead agency also should develop strategies for monitoring, evaluating, and disseminating outcomes.

Education

The agencies of the NNI should continue making investments in innovative and effective education, and the N NCO should consider commissioning a comprehensive evaluation of the outcomes of the overall investment in NNI education.

Societal Impacts

The NSET Subcommittee should develop a clear expectation and strategy for programs in the societal dimensions of nanotechnology. An effective program in societal implications would have well-defined areas of focus, clearly articulated outcomes as well as plans for assessing and evaluating those outcomes, and partnerships that leverage the value of its activities. Ultimately, the inclusion of such programs in the NN I has the goal of streamlining nanotechnology innovation and its positive impact on society, and the creation of new jobs, opportunities and a robust economy.

Nanotechnology Outcomes

The NNAP's review of the NNI's outputs concluded that the NNI has played a critical role in promoting interest and advances in nanotechnology both within the United States and abroad. The United States is clearly the world's leader in nanotechnology R&D and commercialization based on research funding, total number of papers in the most significant scientific publications, patents filed and granted, private sector funding for new and existing companies developing nanotechnologies, and sales of nanotechnology-based products. However, foreign competitors, particularly China, South Korea, Germany, and Japan, are making gains on many of these same metrics. China in particular has significantly increased its share of nanotechnology research publications and patents and now supports nanotechnology as a larger fraction of its total scientific research compared to the United States.

One area in which the United States is unchallenged is in educating nanotechnology researchers. The United States still trains the majority of Ph.D. students in nanoscience and nanotechnology, and though many of these students wish to remain in the United States after completing their degree programs, the data show that over one-third of these students return to their home countries and contribute to the development of nanotechnology R&D programs throughout the world.

To maintain the Nation's leadership role in nanotechnology, the NNAP recommends that the NNI increase its emphasis on nanomanufacturing and commercial deployment of nanotechnology-enabled products, and that the agencies within the NNI must interact and cooperate more with one another to ease the translation of scientific discovery into commercial activity. And though there are benefits to training nanotechnology researchers that return to their home countries, the NNAP also recommends that the Federal Government take steps to retain scientific and engineering talent trained in the United States by developing a program to provide U.S. Permanent Resident Cards for foreign individuals who receive an advanced degree in science or engineering at an accredited institution in the United States and for whom proof of permanent employment in that scientific or engineering discipline exists.

Nanotechnology Outcomes Recommendations

Nanomanufacturing and Commercialization

The National Science Foundation (NSF), Department of Energy (DOE), Department of Defense (DOD), National Institute of Standards and Technology (NIST), and National Institutes of Health (NIH) should include a greater emphasis on manufacturing and commercialization while maintaining or expanding the level of basic research funding in nanotechnology. Specifically, over the next five years, the Federal Government should double the funding devoted to nanomanufacturing (PCA5). In addition, the Federal Government should launch at least five government-industryuniversity partnerships, using the Nanoelectronics Research Initiative as a model. The Federal Government should also support at least five Signature Initiatives over the next two to three years, with each Signature Initiative funded at levels adequate to achieve its stated goals, presumably between $20 million and $40 million annually.

Job Creation

The Department of Commerce and the Small Business Administration should advise the NNI on how to ensure that its programs create new jobs in the United States, including coordinating with State efforts, and economic impact should be an explicit metric in the second decade of the NNI.

Workforce Retention

Congress and the Administration need to take steps to retain scientific and engineering talent trained in the United States by developing a program to provide U.S. Permanent Resident Cards for foreign individuals who receive an advanced degree in science or engineering at an accredited institution in the United States and for whom proof of permanent employment in that scientific or engineering discipline exists.

Moving Nanotechnology to Market

The DOE, DOD, NIST, NIH, National Cancer Institute (NCI), and Food and Drug Administration (FDA) should clarify the development pathway and increase their emphasis on transitioning nanotechnology to commercialization, including making sustained meaningful investments in focused areas to help accelerate technology transfer to the marketplace.

Environment, Health, and Safety

In reviewing the NNI's role in addressing EHS issues, the NNAP concluded that the proactive approach to addressing potential EHS impacts of nanotechnology taken by NNI is commendable. By creating jobs, stimulating economic growth, and providing solutions to some of the toughest challenges facing humankind, nanotechnology has great potential to change the world for the better. Yet realizing this potential may be thwarted if the safety of new materials and products arising from nanotechnology is not addressed up front. In the absence of sound science on the safe use of nanomaterials and of technologies and products containing them, the chance of unintentionally harming people and the environment increases. At the same time, uncertainty and speculation about potential risks threaten to undermine consumer and business confidence.

Over the past two years, the NNI has released a cross-agency nanotechnology EHS research strategy, instigated multi-stakeholder workshops on nanotechnology EHS issues, and seen the Federal nanotechnology EHS research budget increase from $67.9 million in 2008 to a requested $116.9 million in 2011. Individual agencies have also played an active role in international efforts to develop nanotechnology responsibly. Even so, significant EHS-related barriers still stand in the way of effective, sustainable, and responsible commercialization of nanotechnology. As the NNI continues to work toward "a future in which the ability to understand and control matter at the nanoscale leads to a revolution in technology and industry that benefits society," the NNAP recommends that member agencies increase coordinated efforts to overcome these barriers. Specifically, the NSET Subcommittee's interagency working group on Nanotechnology, Environmental, and Health Implications (NEHI) should develop clear principles to support the identification of plausible risks associated with the products of nanotechnology. The NSET Subcommittee's NEHI working group should also further develop and implement a cross-agency strategic plan that links EHS research activities with knowledge gaps and decision-making needs within government and industry to make commercial and regulatory decisions that ensure safe use of nanotechnology products.

The NNAP also recommends that the NSET Subcommittee implement organizational changes that support consequential cross-agency action on addressing nanotechnology EHS issues. In particular, the N NCO should create a senior-level position to lead interagency coordination of efforts in the area of EHS. Finally, the NSET Subcommittee's NEHI working group should develop publicly-available information resources on cross-cutting nanotechnology EHS issues that are relevant to businesses, health and safety professionals, researchers, and consumers.

Environmental, Health, and Safety Recommendations

Risk Identification

The NSET Subcommittee's NEHI working group should develop clear principles to support the identification of plausible risks associated with the products of nanotechnology.

Strategic Planning

The NSET Subcommittee's NEHI working group should further develop and implement a cross- agency strategic plan that links EHS research activities with knowledge gaps and decision-making needs within government and industry.

Information Resources

The NSET Subcommittee's NEHI working group should develop information resources on crosscutting nanotechnology EHS issues that are relevant to businesses, health and safety professionals, researchers, and consumers.

Organizational Changes

The NSET Subcommittee and OSTP should foster administrative changes and communication mechanisms that will enable the NNI to better embrace the EHS issues associated with nanotechnology research, development, and commercialization.

- The NSET Subcommittee co-chairs should assign an individual to NNCO to oversee interagency efforts that address nanotechnology EHS.
- OSTP and the NSET Subcommittee should expand the charter of the NEHI working group to enable the group to address cross-agency nanotechnology-related policy issues more broadly.
- The NSET Subcommittee should explore mechanisms that enable the NSET Subcommittee's NEHI working group to more effectively receive input and advice from nongovernment experts regarding nanotechnology-related risks.

Action Items

The overall conclusion of this review is that the NNI has had a catalytic and substantial impact on the field and should be continued. The NNAP has concluded that the NNI has been, and continues to be, effective in leading and coordinating the United States' effort in nanotechnology. To ensure that the NNI achieves as much if not more in its second decade as it did during its first, the Working Group developed a number of recommendations, listed in the sections above, for consideration by PCAST. While PCAST endorses all of these recommendations, four of these are of special importance. PCAST, in its role as the NNAP, proposes that the President and Congress:

1. Increase funding for the NNI to ensure that the United States retains its leadership role in the development and commercialization of nanotechnology in the face of mounting competition from countries that have responded to the example set by the NNI by investing significant resources in nanotechnology R&D. Over the next five

years, the Federal Government should double the funding devoted to nanomanufacturing, Program Component Area 5 (PCA5). In addition, the Federal Government should launch at least five government-industry-university partnerships, using the Nanoelectronics Research Initiative as a model. The Federal Government should also support at least five Signature Initiatives over the next two to three years, with each Signature Initiative funded at levels adequate to achieve its stated goals, presumably between $20 million and $40 million annually. This funding should be directed primarily to agencies such as the NSF, NIST, DOD, DOE, and NIH. At the same time, the NNI should maintain or expand the level of funding devoted to basic nanotechnology research.

2. Direct the agencies within the NNI to increase the percentage of their nanotechnology related funding provided to the NNCO from $3 million to $5 million, and to require each agency to task senior representatives with decision-making authority to participate in coordination activities of the NNI. The NNCO will use these funds to, among other activities, appoint two individuals to the NNCO to lead interagency coordination of efforts in the areas of EHS research and standards development, respectively. The NNCO must also more actively and aggressively manage the NNI so that it can respond quickly to emerging opportunities and better coordinate interagency efforts, and it must develop metrics for program outputs.
3. Mandate that the NSET Subcommittee's NEHI working group develop a cross-agency strategic plan that links EHS research activities with knowledge gaps and decision-making needs within government and industry, and that the NNCO create a new senior-level position to hold the participating agencies accountable for implementing this strategic plan. This strategic plan must contain clear principles to support the identification of plausible risks based on realistic expectations of exposure to specific nanomaterials.
4. Develop a program to provide U.S. Permanent Resident Cards for foreign individuals who receive an advanced degree in science or engineering at an accredited institution in the United States and for whom proof of permanent employment in that scientific or engineering discipline exists.

A Vision for the Next 10 Years

In its first 10 years, the NNI has promoted progress on scientific problems of great importance to society. It has increased dramatically the investment in instrumentation, research infrastructure, and expertise. Looking forward 10 years, NNAP envisions a program that builds on these strengths, creating new avenues of investigation based on changing societal needs and as yet unforeseen discoveries. A vibrant and effective NNI will have the following attributes:

- ***Basic research will remain a critical component of the research portfolio***. N NI will continue to provide an organizational structure that promotes crosscutting research that enhances our economic competitiveness. New fundamental discoveries will continue to refresh our ideas of what is possible and provide the foundation for new initiatives.

- While basic research continues, there will be increasing focus on integration of components and processes that lead to commercialization. For example, integration of nanotechnologyenabled diagnosis, imaging, and therapy will provide superior new methodologies for developing individualized cancer management strategies.
- The NNI will play a key role in several Signature Initiatives leveraging targeted interagency efforts to address grand challenges. The FY 2011 budget accommodates three grand challenge platforms for nanoscience and nanotechnology: Nanotechnology Applications for Solar Energy, Sustainable Nanomanufacturing, and Nanoelectronics for 2020 and Beyond. These and other bold initiatives will be chosen for their potential to improve quality of life, protect the environment, create jobs, and engage a new generation of scientists and engineers. Among some of the other signature initiatives on which the NN I could partner with industry, academia, and the relevant agencies are regenerative medicine, catalysis, food safety, and threat detection. These initiatives, complemented by a vibrant educational program that focuses on the potential of nanotechnology to help solve societal problems, will excite the imaginations of young people and draw the next generation of scientists and engineers into the field.
- ***The balance of NNI programs will continue to evolve.*** For example, there will be a stronger focus in coming years on fundamental issues related to EHS, an effort that will be coordinated by the NIH, NIST, the Environmental Protection Agency (EPA), and other relevant agencies. Novel means of fabrication with ever finer precision should become the focus of new investments in nanomanufacturing, including modeling and simulation, metrology tools, and the merging of self-assembly with lithography to achieve large scale predictable placement of nanoscale components.

PCAST 2010 National Nanotechnology Initiative Working Group

Co-Chairs

Maxine Savitz*
Vice President
National Academy of Engineering

Ed Penhoet*
Director, Alta Partners
Chairman of the Board, Immune Design
Chairman of the Board, Metabolex

2010 PCAST Nanotechnology Advisory Working Group Members
Peter Antoinette
President and Chief Executive Officer
Nanocomp Technologies, Inc.

Jeffrey Brinker
Laboratory Fellow
Sandia National Laboratory
Distinguished and Regents Professor of
Chemical and Nuclear Engineering and Molecular
Genetics and Microbiology
University of New Mexico

Yet-Ming Chiang
Professor, Dept. of Materials Science and
Engineering
Massachusetts Institute of Technology

Vicki Colvin
Kenneth S. Pitzer-Schlumberger Professor of
Chemistry and Professor of Chemical &
Biomolecular Engineering
Rice University

Mark E. Davis
Warren and Katharine Sch linger Professor of
Chemical Engineering
California Institute of Technology

Garrett Gruener
Co-Founder and Managing Director, Alta
Partners
Chief Executive Officer, Nanomix

Michael Holman
Research Director
Lux Research

Evelyn Hu
Gordon McKay Professor of Applied Physics and
Electrical Engineering
Harvard University

Andrew Maynard
Chief Science Advisor, Project on Emerging
Nanotechnologies
Woodrow Wilson International Center for
Scholars

Chad Mirkin*
George B. Rathmann Professor of Chemistry

Director, International Institute for
Nanotechnology
Northwestern University

Terry Medley
Global Director, Corporate Regulatory Affairs
E.I. duPont de Nemours & Co.

Jennifer Sass
Senior Scientist
Natural Resources Defense Council

Thomas Theis
Director, Physical Sciences
IBM Research, Thomas J. Watson Research Center

Staff

Mary Maxon
Deputy Executive Director, PCAST

Travis Earles
Assistant Director for Nanotechnology
Office of Science and Technology Policy

* PCAST member

I. Introduction and Charge

Chapter Summary

The Federal Government launched the National Nanotechnology Initiative in FY 2001 with a $500 million budget to accelerate the development of nanotechnology. Ten years later, after spending $12 billion of Federal funds and additional private sector funds, the United States is the world leader in both nanotechnology research and development and commercialization. Furthermore, the future looks bright for nanotechnology to become an important contributor to the economic growth of the United States over the coming decade and beyond. Today, the NNI is the Federal Government's crosscutting program that coordinates Federal R&D activities in nanoscale science, engineering, technology, and related efforts among 25 Federal agencies. The NNI is coordinated within the framework of the National Science and Technology Council, the Cabinet-level council by which the President coordinates science and technology policy across the Federal Government. The

National Nanotechnology Coordination Office supports the NSTC Subcommittee as the primary point of contact for the NNI.

The 21st Century Nanotechnology Research and Development Act of 2003 (Public Law 108-153) calls for a National Nanotechnology Advisory Panel to periodically review the Federal nanotechnology R&D program, that is, to review the NNI. The President's Council of Advisors on Science and Technology is designated by Executive Order to serve as the NNAP. PCAST's first review of the NNI was issued in 2005, the second in 2008. This chapter represents the third review of the NNI by PCAST in its capacity as the NNAP. To execute the review, the NNAP formed a 15-person working group. The working group agreed that three overarching categories—program management, nanotechnology outcomes, and environment, health, and safety—would prove most useful for assessing NNI performance and for arriving at valuable and actionable recommendations to ensure that NNI can succeed in the many roles it has to play.

The NNAP recommends that the Administration provide increased support and investment in the NNI as a critical component in the Nation's innovation strategy, with an enhanced focus on commercialization of nanotechnology. Overall, the NNAP has concluded that the NNI has been, and continues to be, effective in leading and coordinating the United States' effort in nanotechnology. No effort this large is perfect, however, and the NNAP has developed a number of recommendations in this document to strengthen and improve the NNI, but the overall conclusion is that the program has had a catalytic and substantive impact on the field and should be continued.

Just over 10 years ago, on January 21, 2000, then President Bill Clinton delivered a speech at the California Institute of Technology in which he adopted the voice of a science-fiction visionary and spoke about nanotechnology:

"Just imagine, materials with 10 times the strength of steel and only a fraction of the weight; shrinking all the information at the Library of Congress into a device the size of a sugar cube; detecting cancerous tumors that are only a few cells in size. Some of our research goals will take 20 or more years to achieve. But that is why, precisely why... there is such a critical role for the Federal Government."

The National Nanotechnology Initiative was launched with an initial investment of $500 million. Now, as the NNI moves into its second decade, and after the Federal Government has spent $12 billion under the NNI rubric, it is important to review the progress that has been made under this bold initiative. It is also important to project the continued potential that nanotechnology offers to society as well as the

What Is Nanotechnology?

Over the years, many people have tried to define nanotechnology. Here is the definition from the NNI Supplement to the President's FY 2011 Budget:[1]

Nanotechnology is the understanding and control of matter at dimensions between approximately 1 and 100 nanometers, where unique phenomena enable novel applications.

Encompassing nanoscale science, engineering, and technology, nanotechnology involves imaging, measuring, modeling, manipulating matter at this length scale.

A nanometer is one-billionth of a meter. A sheet of paper is about 100,000 nanometers thick; a single gold atom is about a third of a nanometer in diameter. Dimensions between approximately 1 and 100 nanometers are known as the nanoscale. Unusual physical, chemical, and biological properties can emerge in materials at the nanoscale. These properties may differ in important ways from the properties of bulk materials and single atoms or molecules.

The promise and optimism about nanotechnology at the launch of the NNI 10 years ago continues to drive many champions and developers of nanotechnology today. Progress has been made in many areas beyond the three examples President Clinton cited in his speech, but nanotechnology is not just a vision of the future. Some forms of nanotechnology are here already, in common use. Consider these examples:

- **Electronics**—The existing semiconductor industry could be described as the most successful and extensive adopter of nanotechnology to date. Current integrated circuits are based on components and structural features in the 30-nanometer range and in some cases even smaller. These dimensions are at least 1000 times smaller than typical biological cells. Every new laptop and iPod works on chips brimming with these nanoscale features.
- **Energy/fuels/environment**—Catalysts and catalytic processes that depend on specific nanoscale structures to steer chemical reactions contribute to a significant portion of the U.S. gross national product (GNP). The catalyst industry and those industries that rely on catalysis exploit this nanotechnology to provide a wide variety of products, such as liquid fuels and plastics, and to contribute to a cleaner environment, such as through the use of catalytic converters to remove pollutants from automobile exhaust. Additionally, materials for high-power, fast-charging batteries used in many cordless power tools incorporate advanced electrodes whose capabilities depend on deliberately-engineered nanoscale architectures.
- **Medicine**—Several nanoparticulate formulations of conventional drugs are being used in the treatment of cancer and infectious disease. A number of nanotechnology-based imaging agents and therapeutics that target tumor cells and arterial plaques are in clinical trials. In addition, nanotechnology-based detectors form the core of a number of new diagnostic instruments that are better than previous generations of instruments at detecting minute quantities of important biomarkers of disease.
- **Materials**—Carbon nanotubes are currently being incorporated into high-strength composites and woven into yarns to produce significantly lighter and more conductive wires and electrical harnesses.
- **Consumer products**—Nanoscale materials and particles are being used increasingly as ingredients in cosmetics, sunscreens, and food products. The small sizes of the particles confer various properties, such as high sun-blocking power with translucency in sunscreens, stain resistance for fabrics, and self-cleaning properties and better color features for paints.

The National Nanotechnology Initiative

The NNI is the U.S. Government's crosscutting program that coordinates Federal research and development activities in nanoscale science, engineering, technology, and related efforts among various participating agencies. The NNI was established in FY 2001 with an initial budget of $500 million and with six Federal agencies as original participants. Today, the NN I is a multi-agency initiative that includes 25 Federal agencies, 15 of which in FY 2011 have their own individual budgets for nanotechnology R&D. Overall, the budget number and allocation among the NNI's Program Component Areas is determined in a collaborative manner among OSTP, OMB, and the participating agencies. In FY 2010, the budget for 13 agencies totals $1.6 billion. The FY 2011 budget request totals $1.8 billion, and includes increased R&D investments in two PCAs: Nanomanufacturing and Environment, Health, and Safety. The budget request also calls for FDA and the Consumer Products Safety Commission (CPSC) to join the roster of agencies that fund nanotechnology R&D. The cumulative investment in nanotechnology by the Federal Government since the NNI's inception, including the $1.8 billion requested for FY 2011, totals about $14 billion.

The N NI is managed within the framework of the National Science and Technology Council, the Cabinet-level council by which the President coordinates science and technology policy across the Federal Government. The NSET Subcommittee of the NSTC coordinates the planning, budgeting, program implementation, and review of the NN I. The NSET Subcommittee is composed of representatives from Federal agencies participating in the NNI (see Appendix E).

The NNCO provides technical and administrative support to the NSET Subcommittee, serves as a central point of contact for Federal nanotechnology R&D activities, and engages in public outreach on behalf of the NNI. The NNCO also serves as a liaison to academia, industry, professional societies, foreign organizations, and others to exchange technical and programmatic information. Additionally, the NNCO coordinates preparation and publication of NNI interagency planning, budget, and assessment documents, and maintains the NNI Web site, www.nano.gov.

The December 2007 N NI Strategic Plan[2] reiterated the vision for the N NI: a future in which the ability to understand and control matter at the nanoscale leads to a revolution in technology and industry that benefits society. The plan specifies the following four goals to achieve the overall vision:

1. Advance a world-class nanotechnology research and development program;
2. Foster the transfer of new technologies into products for commercial and public benefit;
3. Develop and sustain educational resources, a skilled workforce, and the supporting infrastructure and tools to advance nanotechnology; and
4. Support responsible development of nanotechnology that considers the technology's environmental, health, safety, and broader societal dimensions.

National Nanotechnology Advisory Panel

The 21st Century Nanotechnology Research and Development Act of 2003 (Public Law 108-153) calls for a National Nanotechnology Advisory Panel to periodically review the Federal nanotechnology R&D program, that is, to review the NNI. The President's Council of Advisors on Science and Technology is designated by Executive Order to serve as the N NAP. Section 4 of the law states that the Advisory Panel shall advise the President and PCAST on matters relating to the NNI (referred to also as "the program"), including assessing:

- Trends and developments in nanotechnology science and engineering;
- Progress made in implementing the program;
- The need to revise the program;
- The balance among the components of the program, including funding levels for the program component areas;
- Whether the program component areas, priorities, and technical goals developed by the Council are helping to maintain U.S. leadership in nanotechnology;
- The management, coordination, implementation, and activities of the program; and
- Whether societal, ethical, legal, environmental, and workforce concerns are adequately addressed by the program.

NNAP's first review of the NNI was issued in 2005;[3] the second one was issued in 2008.[4] A summary statement in the 2008 report about the United States' status in the global nanotechnology R&D movement, which could be an indication of NNI's performance, included the following assessment:

"The United States remains a leader in nanotechnology based on various metrics, including R&D expenditures and outputs such as publications, citations, and patents. However, the European Union has more publications and China's output is increasing."

Regarding the NNI's role and performance in this context, the NNAP stated in their 2008 report, among other things, that:

- The ongoing NNI investment in infrastructure and instrumentation is commended and encouraged;
- Advances in nanotechnology are embodied in a growing number of applications in various industries;
- The approach for addressing EHS research under the NNI is sound;
- In consultation with the President's Council on Bioethics, the panel concluded that at present, nanotechnology does not raise ethical concerns that are unique to the field; and
- NNI continues to be a highly successful model for an interagency program; it is well organized and well managed.

Based on its assessment, the prior NNAP made six recommendations in 2008 to strengthen the NNI. They are paraphrased below:

1. **Infrastructure, management, and coordination**—The NNI should improve the intra-agency coordination, and NNI member agencies should support international coordination through international forums such as the Organisation for Economic Co-operation and Development (OECD).
2. **Standards development**—Federal agencies should continue to engage in national and international standards development activities. A strong U.S. representation should be maintained in international forums and efforts made to avoid duplicative standards work. Where appropriate, NIST and other NNI agencies should develop reference materials, test methods, and other standards that provide broad support for industry production of safe nanotechnology-based products.
3. **Technology transfer and commercialization**—NNI funded centers should be structured to encourage partnering with industry. The NNI should seek means to assess more accurately nanotechnology-related innovation and commercialization of NNI research results
4. **Environmental, health, and safety implications**—EHS research for nanomaterials should be coordinated with those taking place in industry and with programs funded by other governments to avoid gaps and unnecessary duplication. EHS research should be coordinated with, not segregated from, applications research to promote the coordinated consideration of risks and benefits associated with nanomaterials and technologies incorporating them.
5. **Societal and ethical implications**—Research on the societal and ethical aspects of nano- technology should be integrated with technical R&D and take place in the context of broader societal and ethical scholarship.
6. **Communication and outreach**—The NNI should expand outreach and communication activities by the NNCO and the Nanotechnology Public Engagement and Communications Working Group and by coordinating existing agency communication efforts. Information should be developed with broad input and should be made available in ways that incorporate two-way communication.

In addition to the two prior NNAP reviews, the NNI is subject to periodic reviews by other Federal and non-Federal panels with appropriate expertise. For example, the 21st Century Nanotechnology Research and Development Act of 2003 directs the National Research Council to conduct a triennial external review of the NNI. The first NRC review was published in 2006. The second triennial review has not been initiated, but it would be useful to have the detailed program review conducted by the NRC to complement this review by the NNAP.

In 2009, the NRC also published a more specific review of the NNI's EHS program, which is discussed in chapter four of this document. As this third NNAP review was being prepared, the NRC was also conducting another EHS study titled "Research Strategy for Environmental, Health, and Safety Aspects of Engineered Nanomaterials."

The GAO currently has a review underway that is examining the current and future uses for manufactured nanomaterials, what is known about potential risks of these materials, and EPA's efforts to understand and regulate these risks. This work is expected to be released in the summer of 2010.

The 2010 Review of NNI by the NNAP

This chapter represents the third review of the NNI by PCAST in its capacity as the NNAP. To execute the review, the NNAP formed a 15-person working group. The co-chairs and one other member of the working group are members of PCAST. The other members of the working group consisted of external representatives from academia, industry, and nongovernmental organizations (NGOs). In developing the Statement of Task (SOT) and related questions for this review (Appendix A), the co-chairs of the working group received input from the other members of PCAST, relevant Congressional committee staff, and staff from GAO, OMB, NRC, and OSTP (Appendix B).

Two in-person working group meetings were held for preparatory purposes, to obtain input from technical experts (Appendix C) from government, NGOs, academia, and industry; to gather information and discuss issues; and to develop a range of recommendations that were considered by the N NAP. For additional input, experts were asked to answer a series of questions (Appendix B) related to the seven areas of focus defined by 21st Century Nanotechnology Research and Development Act, as well as a number of additional questions developed by the working group. This process is described in Appendix D.

The NNI review group agreed that there were three overarching categories that would prove most useful for assessing NNI performance and for arriving at valuable and actionable recommendations to ensure that NNI can succeed in the many roles it has to play. Those categories are:

1. **Program management**—An appraisal of how well NNI leadership has performed with respect to the roles it has been tasked to carry out.
2. **Nanotechnology outcomes**—An analysis of what the Federal nano investment has delivered and recommendations to enhance the outcomes, especially economic outcomes.
3. **Environment, health, and safety**—An assessment of N NI's performance in helping to orchestrate the identification and management of potential risks associated with nanotechnology, with particular attention paid to reviewing progress the NNI has made in following through on recommendations made in the 2008 NNAP review of the NNI.

Each of these categories forms the basis of the three chapters that follow. Because 2010 marks the 10th anniversary of the NNI, the panel decided it would be appropriate to conclude with a forward-looking chapter that discusses how nanotechnology might contribute to important societal needs and goals in the coming years.

There are a number of recommendations in this document to strengthen and improve the NNI, but the overall conclusion is that the program has had a catalytic and substantial impact on the field and should be continued. The N NAP recommends that the Administration, through OSTP and OMB, provide increased support and investment in the NNI as a critical component in the nation's innovation strategy with an enhanced focus on commercialization of nanotechnology.

II. Program Management

Chapter Summary

Over the course of its first decade, the NNI has distinguished itself as a successful cooperative venture involving the participation of 25 Federal agencies. The NNAP believes that the NNI has been well organized and managed, but that there are several steps that OSTP can take to enable the NNI over the next decade to fully exploit the opportunities offered by the development of nanoscience and nanotechnology, particularly in regards to commercialization efforts.

In particular, the NNAP believes that the NNI would benefit by OSTP designating more resources for the NNCO to coordinate activities across the NNI and by authorizing dedicated funding of approximately $5 million annually (0.3 percent of agency N NI funding, up from less than 0.2 percent of agency NNI funding, or almost $3 million, in FY 2010) to support that mission. The NNCO will use these funds to, among other activities, appoint two individuals to the NNCO to lead interagency coordination of efforts in the areas of EHS research and standards development, respectively. The NNCO should also conduct regular reassessments of the balance of funding to each of the PCAs. The NNCO should also continue to spearhead efforts in nanotechnology education and develop an overall strategy to guide research on the societal impacts of nanoscience and nanotechnology that include, but also go beyond, EHS issues.

From its inception, the NNI has been distinguished by broad interagency participation that can collectively span the breadth of investments needed to fully exploit the opportunities offered by the development of nanoscience and nanotechnology. Beginning with six agencies at its establishment in 2001, the NNI currently involves the nanotechnology-related activities of 25 Federal agencies, 15 of which have specific budgets to fund R&D in this area.[5] The NNCO acts as the primary point of contact for information on the NNI and provides technical and administrative support for the NSET Subcommittee, the interagency body that coordinates planning, budgeting, and program implementation for the NNI. As a focal point of information about the NNI, the NNCO maintains the NNI website and provides public outreach on behalf of the NNI. At present, the NNCO operates with one Federal employee serving as director via an NSTC agency detail and a staff of six to seven contractors. The current NNCO budget represents less than 0.2 percent (under $3 million in FY 2010) of the NNI total budget and comes from contributions provided voluntarily by the participating funding agencies.

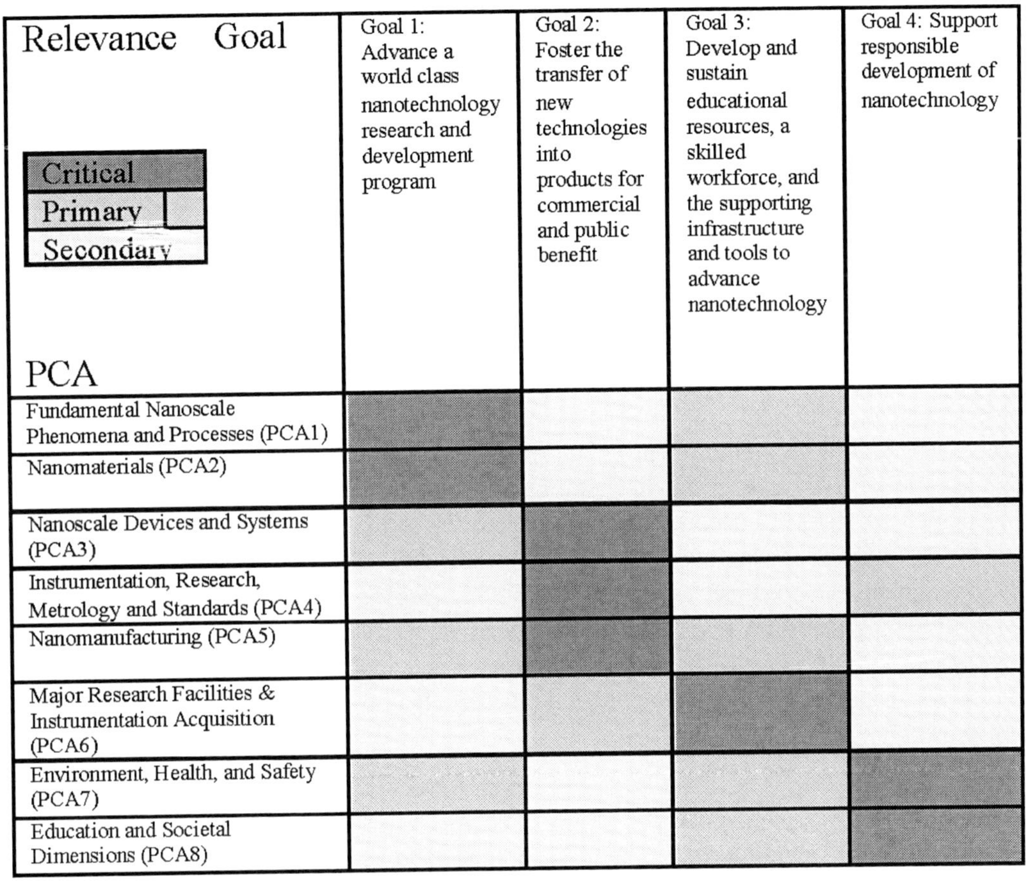

Source: NNI 2007 Strategic Plan

Figue 2-1. Relationships between the picas and NNI goals

In addition to the four primary goals of the NNI listed in this review's Introduction, the program revised its strategic plan in 2007 to include the following eight specific Program Component Areas:

1. Fundamental Nanoscale Phenomena and Processes
2. Nanomaterials
3. Nanoscale Devices and Systems
4. Instrumentation Research, Metrology, and Standards for Nanotechnology
5. Nanomanufacturing
6. Major Research Facilities and Instrumentation Acquisition
7. Environment, Health, and Safety
8. Education and Societal Dimensions

The PCAs provide an organizational framework for the activities of the NN I and serve to group together related projects and activities. The relationship between the PCAs and NNI goals can be represented by Figure 2-1.

Overall, the NNAP believes that the NNI continues to serve as a commendable model of an interagency program that is generally well organized and managed. The goals of the N NI are comprehensive and cogent, the PCAs remain an effective means of tracking the investments in the NNI, and the working groups[6] of the NSET Subcommittee appropriately provide added interagency focus on crosscutting topics of importance to the N NI and the realization of its goals. However, the NNAP believes that there are ways in which the program management could be further strengthened as the NN I enters its second decade.

Issues for a Maturing NNI: Beyond Year 10

The NNI is in its 1 0th year of Federal funding, having garnered bipartisan support in an exceptional, long-term commitment to innovation and transformative change. As the NN I matures, and with a range of nanotechnologies primed to move into commercialization and implementation, the need becomes even greater for close interagency coordination, for well-conceived strategies for investments, and for the identification of measurable outcomes for that investment. Also critical is the effective collection and dissemination of information about NNI programs and resources to the broad range of stakeholders. The NNAP thus expects that the NNCO will have an even greater and more critical role to play in communications and coordination among the various participants of the NNI.

RECOMMENDATION 2-1. STRENGTHEN THE NNCO

The NNCO should broaden its impact and efficacy and improve its ability to coordinate and develop NNI programs and policies related to those programs. OSTP should facilitate these improvements by taking the following actions:

- Require each agency in the NNI to have senior representatives with decision-making authority participate in coordination activities of the NNI.
- Strengthen the NNCO to enhance its ability to act as the coordinating entity for the NNI.
- Mandate that the NNCO develop metrics for program outputs and that it works with the Bureau of Economic Analysis to develop metrics and to collect data on the economic impacts of the NNI.
- Appoint two individuals to the NNCO to lead interagency coordination of efforts in the areas of EHS research and standards development, respectively.
- Dedicate 0.3 percent of NNI funding to the NNCO to ensure the appropriate staffing and budget to effectively develop, monitor and assess NNI programs.

To strengthen the decision-making process and implementation of coordinated NNI activities, each NNI member agency should designate senior representatives with decision-making authority to participate in the NSET Subcommittee. With a stronger coordinating body that has the knowledge and authority to formulate and monitor programs, the NNI will be better prepared to take on new strategic initiatives, fine tune existing programs, allocate investments in ways that are more responsive to new opportunities and needs, and serve as a communications hub among NNI participants as well as for the public and all stakeholders.

A strengthened NNCO should be the coordinating agency to ensure several desirable improvements in managing the NNI, including:

- Enhancing communications with the business sector;
- Facilitating technology transfer;
- Providing information on available nanotechnology resources to both the public and private sectors. The NNI has created outstanding resources for the nanotechnology R&D community through its investments in infrastructure such as shared user facilities, research centers and networks, and education centers and networks. Accessible information on these facilities and coordination of their use would ensure optimal leverage of these considerable investments;
- Collaborating with stakeholders on enabling programs such as metrology; standards including size, shape and composition of nanomaterials, and databases of physical and chemical properties of nanomaterials; and manufacturing safety; and
- Engaging in closer and more frequent interactions with States, which could provide important leverage of resources for the NNI. The workshops previously held on regional, state, and local initiatives are commendable, but these opportunities should be provided more often for more active ongoing engagement.

In addition, the NNAP believes that as programs are developed within the NNI, the NNCO should track relevant metrics to measure the outcomes and impacts of NNI programs. To assess systematically the outcomes of NNI investment, it would be desirable, for example, to be able to measure the value of all products with nanotechnology components, as well as the value of those components. Following both of these measures over time would give a reasonable picture of the importance of nanotechnology in the economy. Such measurements would in turn enable an estimate of the number of jobs created by nanotechnology, as well as the social rate of return to NNI investment. Rather than relying on funding agencies for such estimates, it would be most appropriate to lodge responsibility with a statistical agency such as the Bureau of Economic Analysis in the Department of Commerce.

The NNCO needs to monitor metrics that assess the NNI's impact, such as on job creation, commercial activity, private sector investment including venture funding, advanced degrees, international competitiveness, and the development of methodology for assessing plausible risks associated with nanotechnology.

To accomplish the range of tasks described above, the NNAP recommends that OSTP increase the set- aside from the NNI budget allocated to the NNCO to 0.3 percent of NNI funding, or about $5 million annually.

Figure 2-2 tracks the relative NNI investment per PCA from 2006 (when they were initially established) to 201 0. The largest percentage investment has been in PCA1 (Fundamental Nanoscale Phenomena). The combined funding for PCA1 and PCA2 (Nanomaterials) has comprised about 50 percent of the total NNI funding; this reflects the importance of investments for fundamental research in nanoscience and nano- technology, which will underlie further innovation, applications, and potential for commercialization.

RECOMMENDATION 2-2. FOCUS ON COMMERCIALIZATION

In a budget planning process coordinated by OSTP, each agency should continually re-evaluate its NNI balance of investments among the PCAs, with an enhanced focus on commercialization, which would include maintaining the current level of investment in research and doubling the investment in nanomanufacturing (PCA5) over the next five years.

Although the amount of requested NNI funding changed from $1 billion in 2006 to $1.64 billion in 2010, Figure 2-2 shows that there has been little change in the distribution among the PCAs over those five years. There has been a small change in distribution in the FY 2011 budget request, allowing for significant increases in Nanomanufacturing (PCA5) of $26 million, or about 34 percent more than 2009. Funding for EHS research (PCA7) increased from $35 million in 2006 to $1 17 million in the FY 2011 request. Examination of Figure 2-2 suggests that these increased investments were largely offset by a decrease in investments devoted to PCA1 ; the overall distribution of funds to other PCAs remains similar to those in previous years.

The re-examination of the FY 2011 budget and the increased emphasis on nanomanufacturing is laudable. However, it is important to realize that as the programs of the NNI mature, the appropriate percentage investments into the PCAs would be expected to evolve. For example, one might also expect increasing emphasis on Nanoscale Devices and Systems (PCA3) and Metrology and Standards (PCA4), and that different kinds of Major Research Facilities (PCA6) may be needed and appropriate. A continual re-evaluation of investments among all the PCAs is thus important: an increase in funding of nanomanufacturing alone will not provide the full complement of resources required to usher the fundamental R&D fruits of the NNI into phases of commercialization and implementation.

In addition, agencies with Small Business Innovation Research (SBIR) and Small Business Technology Transfer (STTR) programs should include the PCAs in their planning processes. The NNCO should work with the senior management from each agency to develop the full N NI funding profile, thereby leveraging existing resources and preventing duplications of efforts.

Nanotechnology Signature Initiatives in the Service of Society

The NNAP applauds the introduction of the three Signature Initiatives described in the FY 2011 budget. The NNAP views these three grand challenge platforms for nanoscience and nanotechnology— Nanotechnology Applications for Solar Energy, Sustainable Nanomanufacturing, and Nanoelectronics for 2020 and Beyond—as new opportunities for collaborative involvement of the participating agencies

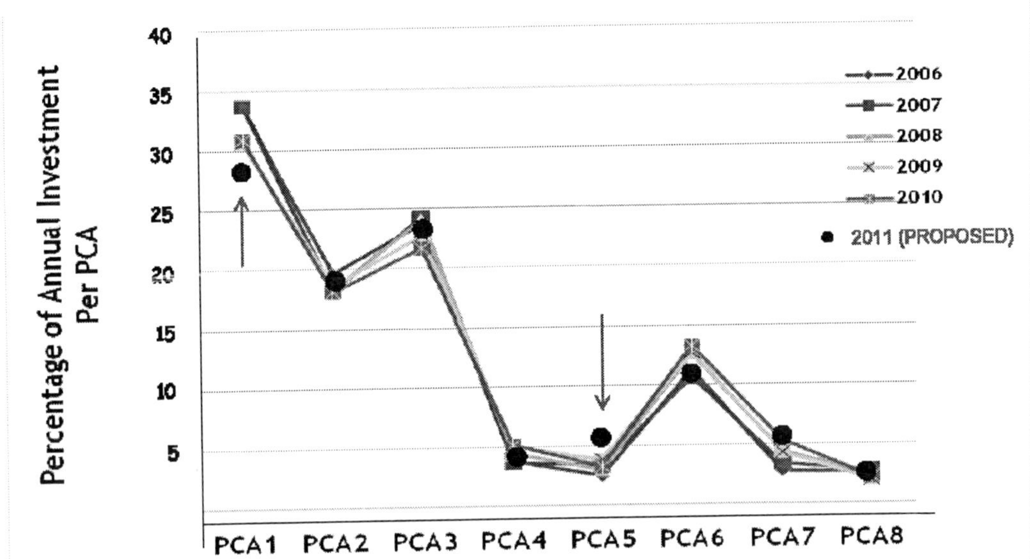

Source: C. Teague, Survey (2006-2010) and NNI – Supplement to the President's 2011 Budget. Red arrows indicate changes in allocation of the 2011 proposed budget.

Figure 2-2. Percentages of Annual NNI Investment Per PCA, 2006-2011

RECOMMENDATION 2-3. SIGNATURE INITIATIVES

Each Signature Initiative's lead agency should develop coordinated milestones, promote strong educational components, and create public-private partnerships to leverage the outcomes of the initiatives. Each lead agency also should develop strategies for monitoring, evaluating, and disseminating outcomes.

The NNAP also believes that these initiatives will inspire and attract outstanding students by providing them with opportunities to apply leading edge nanoscience and nanotechnology in the service of society. The success of these Signature Initiatives is particularly important in light of the challenges represented by well-funded and increasingly aggressive overseas nanotechnology programs.

New Knowledge, New Themes, and New Means of Learning

The NNI has distinguished itself in its focus on education and outreach. In particular, funding for National Science and Engineering Centers (NSECs) and Materials Research Science and Engineering Centers (MRSECs) has provided an unusually rich, multi-disciplinary educational environment intrinsic to nanoscience and nanotechnology. The National Center for Learning and Teaching in Nanoscale Science has provided an outstanding resource for students and teachers alike, and the Nanoscale Informal Science Education (NISE) Network has done much to underscore the tremendous efficacy of interactive displays, various forms of dialogue and engagement between scientists and the public, and other informal learning situations.

RECOMMENDATION 2-4. EDUCATION

The agencies of the NNI should continue making investments in innovative and effective education, and the NNCO should consider commissioning a comprehensive evaluation of the outcomes of the overall investment in NNI education.

Either through the NNI or independently, various master's and Ph.D. programs have been established in nanotechnology. A well-educated workforce and public are critical to the continued prosperity of our society, and these programs should be continued from grades K-12 and beyond. The NNAP urges that the agencies in the N NI continue to seek creative programs that engage and educate broad sectors of the public. The NNAP also recommends strongly that a comprehensive review of such programs be undertaken to better understand what outcomes and benefits have been realized.

Programs in Societal Implications

The inclusion of programs involving the societal implications of nanotechnology, both as part of center- funded initiatives as well as through specifically established Nanotechnology and Society Centers, has been a laudable signature of the NNI. Studies on the societal dimensions of nanotechnology can encompass a broad spectrum of issues: such studies can involve the public perception of research in nanotechnology and its attendant benefits and possible risks, the economic and global aspects of the nanotechnology enterprise, and achieving an understanding of the process of innovation and the path to commercialization.

As nanotechnology is finding increasing application in a wide range of products and contexts, it is critical that the NNCO develops a clear strategy and focus for this program. A strategic and effective investment in the societal aspects of nanotechnology would define which areas of research to focus on, what outcomes are expected, what partnerships would best guarantee desired outcomes, and how those outcomes are to be evaluated.

RECOMMENDATION 2-5. SOCIETAL IMPACTS

The NSET Subcommittee should develop a clear expectation and strategy for programs in the societal dimensions of nanotechnology. An effective program in societal implications would have well-defined areas of focus, clearly articulated outcomes as well as plans for assessing and evaluating those outcomes, and partnerships that leverage the value of its activities. Ultimately, the inclusion of such programs in the NNI has the goal of streamlining nanotechnology innovation and its positive impact on society, and the creation of new jobs, opportunities and a robust economy.

III. Outputs of Federal Nanotechnology Research

Chapter Summary

In this chapter, the Working Group reviews the output for the NNI and that of other countries that are developing their own national nanotechnology efforts. This review shows that the NNI has played a critical role in promoting interest and advances in nanotechnology both within the United States and abroad. Though still the leader in nanotechnology R&D and commercialization, the United States is losing ground to foreign competitors, particularly China, South Korea, Germany, and Japan, on a number of key metrics of research output and commercial activity. The United States still trains the majority of Ph.D. students in nanoscience and nanotechnology, but more than one-third of these students do not stay in the United States, even when, in many cases, they wish to stay and when U.S. companies or universities wish to retain them. Having the world's top talent has been critical for the United States' success in nanotechnology, and if this talent continues tomove overseas, the economic benefits and job creation will follow."

To maintain our leadership role in nanotechnology, the NNAP recommends that the NNI increase its emphasis on nanoman ufacturing and commercial deployment of nanotechnology-enabled products, and that the agencies within the NNI must interact and cooperate more with one another to ease the translation of scientific discovery into commercial activity. The NNAP also recommends that the Federal Government take steps to retain scientific and engineering talent trained in the United States by developing a program to provide U.S. Permanent Resident Cards for foreign individuals who receive an advanced degree in science or engineering at an accredited institution in the United States and for whom proof of permanent employment in that scientific or engineering discipline exists.

In the 10 years since the FY 2001 launch of the NNI, worldwide interest and investment in nanotechnology R&D have soared. Today, virtually every country that supports scientific R&D has a nanotechnology initiative, and some nations are on a path that could lead to serious challenges to the leadership of the United States in the development of nanotechnology-based products. This chapter examines how the NNI is affecting the development of innovative products within the United States and in comparison to the rest of the world, with an eye on assessing how the United States is faring in what is becoming heated competition to reap the fruits of nanotechnology R&D. This chapter also considers current barriers that hinder the transfer of research advances into the marketplace.

Trends and Developments in Nanotechnology Science & Engineering

The NNAP has observed a number of general trends since its last review of the NNI in 2008. Nanotechnology science and engineering is becoming more interdisciplinary and global, with innovation being particularly strong at the intersection of fields. The fundamental science base built over the past decade is today yielding a rapidly growing number of nanotechnology-based products and companies. This growth has been accompanied by an increasing emphasis on the infrastructure needed to support the commercialization of nanotechnologies and to study potential societal impacts, most notably in the areas of EHS and nanomanufacturing. Matching pace with this growth, funding as a metric of the level of nanotechnology activity has continued to increase as well, with total global funding in 2008 increasing 15 percent to $18.2 billion from $15.8 billion in 2007. The United States' share of the 2008 total was $5.7 billion, up 9 percent from the $5.2 billion that the government, corporations, and venture capitalists invested in nanotechnology in 2007. Government funding in 2008 was 46 percent of total funding, an increase of 16 percent from 2007 levels.[7]

In 2001, the N NI's main focus was to revolutionize the development of microelectronics and new manufacturing technologies. There have been notable successes in this area, such as the development and widespread use of high-resolution lithography. However, new opportunities have also arisen in areas that were nascent or unanticipated at the inception of NNI 10 years ago, and that today could be the focus of "signature initiatives" (some of which are included in the FY 2011 budget request) that would drive the next 10 years of the NNI. These include:

- Nanomedicine (e.g., imaging, diagnostics and therapeutics; antibiotics; gene regulation);
- Nnergy storage, generation, and conversion;
- Environmental diagnostics and cleanup;
- Structural and multi-functional nanocomposites;
- Homeland security (e.g., chemical and biological detection and monitoring); and
- Manufacturing nanotechnology-based products.

During the last several years, basic research on carbon nanotubes, composites, and coatings has declined even as applications of nanotechnology have grown. For example, several companies today are commercializing low-cost carbon nanotubes for use as conductive components, such as in the electrodes in advanced batteries (a company developing novel nanotube-based electrical composites is highlighted later in this chapter). Research on materials such as graphene, metamaterials, silicon nanowire thermoelectrics, and plasmon-enhanced solar cells has increased. About one-third of current nanotechnology research is related to sensors, and nanotechnology research in the life sciences has doubled since 2002.

Moreover, it is recognized that nanotechnology can play a critical role in addressing global problems of energy supply, climate change, and sustainability. Energy and environmental applications of nano- technology have undergone explosive growth. In 2008, this sector accounted for 29 percent of all nano- technology funding by the Federal Government, 14 percent of all corporate nanotechnology funding, 41 percent of venture capital funding, 21 percent of nanotechnology publications, and 59 percent of all nanotechnology patents.[8]

Since the last NNAP review, a number of commercialization landmarks occurred in nanotechnology:

- The private sector has continued to invest in nanotechnology. Worldwide corporate funding of nanotechnology R&D first exceeded government funding in 2007, and this trend continued in 2008; in the United States, private funding (including corporate R&D and venture capital investments) was more than double the total of Federal nanotechnology funding.[9]
- An estimated $224 billion worth of *final products* sold worldwide made some use of nanotechnology components in 2009, up from $135 billion in 2007. Products made in the United States accounted for $80 billion of the final products, though the share of the global sales attributed to the United States declined from 39 percent to 35 percent.[10]
- The essential nanotechnology *components* (nanointermediates) that went into these products amounted to $29 billion worth of sales; $11 billion worth of these components were made in the United States.
- Among these products, the largest share, 55 percent, was from the materials and manufacturing industry sector, including products such as automobiles, industrial equipment, and building and construction. These products made use of nanotechnology-based components such as coatings, composites, and electronic components. Next, 32 percent of the final products were from the electronics and information technology sector, and included items such as mobile devices with displays and antimicrobial coatings enhanced with nanotechnology. Some 12 percent of nanotechnology-based products were in the healthcare and life sciences sector, primarily from nano-enabled drug delivery systems, while 1 percent of nanotechnology-based products came from the energy and environmental sector and included items such as nano-enabled filtration membranes or batteries.

At the level of individual States, nanotechnology is vibrant and active. Today, some 25 States have their own nanotechnology programs. Most State efforts leverage Federal N NI-supported research by emphasizing translational research and development aimed at state and regional job creation by the private sector. Among the most ambitious of these is the College of Nanoscale Science and Engineering at the State University of New York, Albany. With New York State assistance, the College has built infrastructure consisting of 800,000 square feet of shared new facilities for public-private partnerships. More than a dozen companies involved in the development of nanoelectronics technologies now use this facility.

Assessing U.S. Leadership in Nanotech nology

While the United States continues to lead in most areas, it has lost ground to foreign competitors based on several key metrics. Applicable metrics include the number of scientific publications, citations to published literature, patents, the amount of government and corporate spending, the number of nano- technology centers and initiatives, the number of Ph.D. graduates, and the number of active companies. The rate of growth in certain of these is especially high for China, South Korea, Germany, and Japan, as this chapter explains in more detail below.

Research activity as measured by publications and citations

At the time of the last N NAP review in 2008, the United States had lost ground to the 27 nations in the European Union (EU27) in total number of nanotechnology publications[11] and was nearly equal to China (including Taiwan), as shown in Figure 3-1 . Since then, the total number of publications from U.S. laboratories has decreased slightly, while the number of Chinese publications has continued to climb. As a result, in 2009 the United States is now third behind China and the EU27 in the total number of nanotechnology publications.

The total number of publications does not necessarily reflect quality or influence, however. Many of the publications from China are not published in the 12 core journals of nanoscience and nanotechnology,[12] while researchers from the United States and the EU27 continue to contribute the majority of papers appearing in those journals. Nonetheless, China's share of publications in these journals is increasing at about the same rate as the United States' share is decreasing, as shown in Figure 3-2.

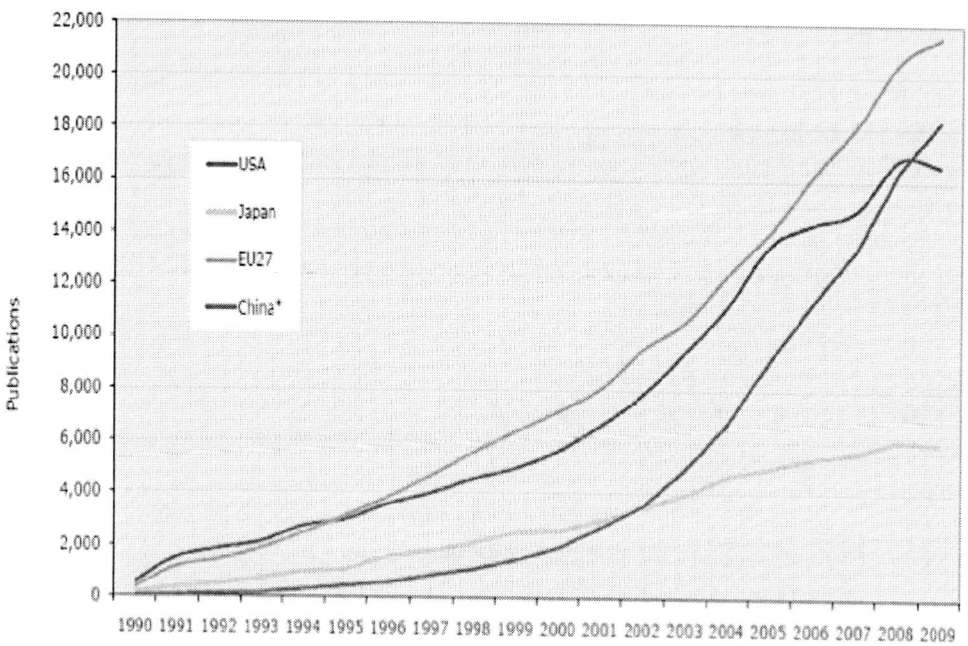

Source: Chen HC, Dang M, Roco MC. "Updated Nanotechnology Indicators, January 201 0." Addendum to Chen H, Roco MC (eds). *Mapping Nanotechnology Innovations and Knowledge: Global, Longitudinal Patent and Literature Analysis*. New York: Springer-Verlag, 2008.

Figure 3-1. Total number of nanotechnology publications appearing in the Science Citation Index by year. Numbers for China include Taiwan.

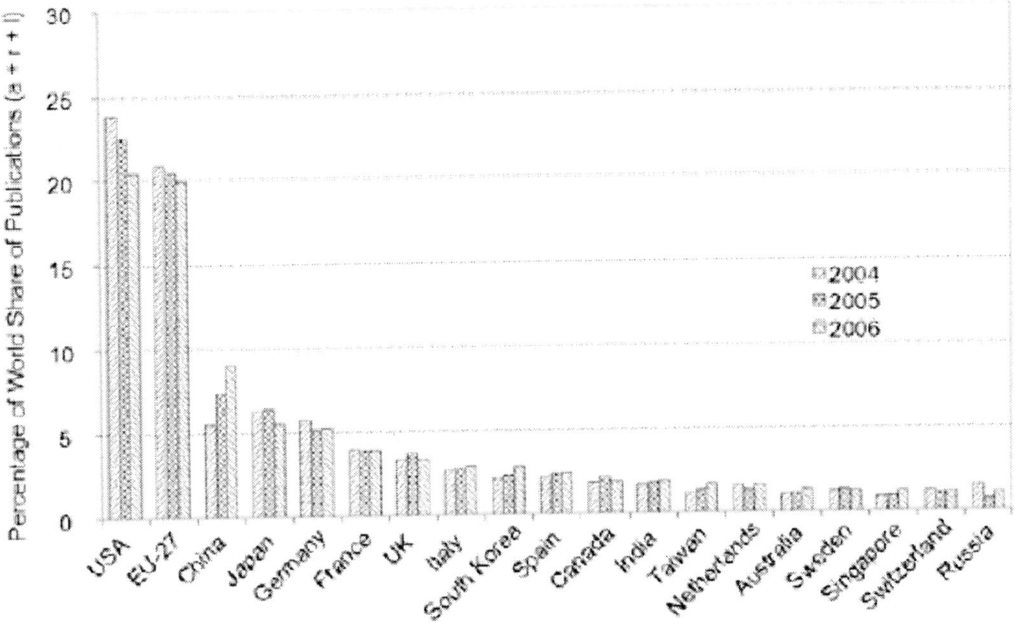

Source: Leydesdorif L, The delineation of nanoscience and nanotechnology in terms of journals and patents: A most recent update, Scientometrics, 2008; 76(1):1 59–167

Figure 3-2. Percentage of world share of publications in 12 core journals of nanoscience and nanotechnology by country, for the years 2004-2006.

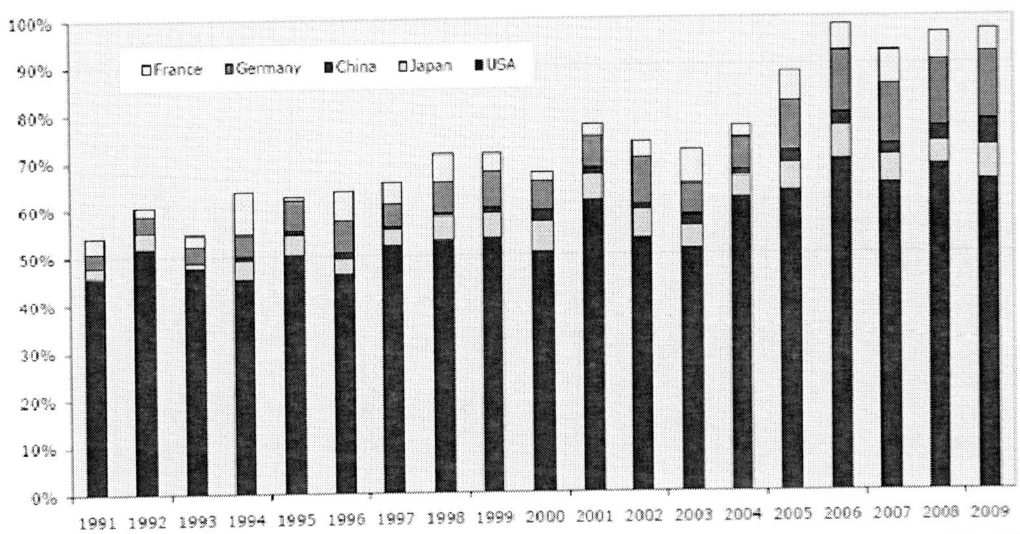

Source: Chen HC, Dang M, Roco MC. "Updated Nanotechnology Indicators, January 201 0." Addendum to Chen H, Roco MC (eds). *Mapping Nanotechnology Innovations and Knowledge: Global, Longitudinal Patent and Literature Analysis*. New York: Springer-Verlag, 2008.

Figure 3-3. Percentage of nanotechnology papers published in *Science*, *Nature*, and *Proceedings of the National Academy of Sciences* that are cited by other publications and patents, organized by country of the research institution.

As another measure of publication quality, consider publications in the three journals *Science, Nature, and Proceedings of the National Academy of Sciences*; papers appearing in these three journals are most often cited in other nanotechnology papers and patents, a measuring of scientific importance. According to this metric, the United States continues to hold a dominant position with about 65 percent of the total number of citations, as shown in Figure 3-3. Nonetheless, the trend since 2005 shows that the percentage of such citations attributed to the United States, France, Germany, and Japan are relatively static, while China's share increased sharply in 2008–2009 and may soon exceed that of France and Japan.

Thus, while the United States remains the leader in scientific publication quality, it no longer dominates in total nanotechnology research output. Moreover, publication trends for other countries indicate a proportionally greater emphasis on nanotechnology. In China, Japan, South Korea, India, Taiwan, and Singapore, nanotechnology research is a larger fraction of all scientific research than it is in the United States.

Patents

Patent activity is a key metric of technology creation. The United States remains the world leader by a large margin in the absolute number of nanotechnology patents *issued*, with more than 1500 nanotechnology patents being issued in each of 2007 and 2008. The total number of nanotechnology patents issued to U.S. inventors since 1995 exceeds 10,000.[13] Germany and Japan are a distant second and third with about 10 times fewer patents issued. However, the number of patents issued is a lagging indicator, while the number of patent applications *filed* is a forward indicator as it typically takes several years for filed patents to issue. As shown in Figure 3-4, patent filing activity in China rose sharply during the period 2000-2004 and overtook the United States during 2005-2008.

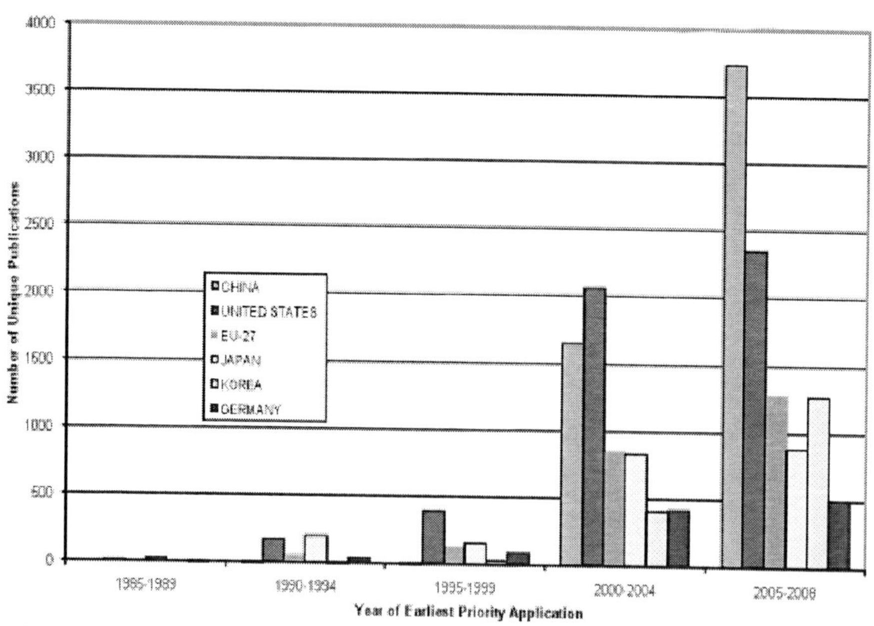

Source: Kisliuk B, USTPO, unpublished study on comparative patent filings, January 2010

Figure 3-4. Nanotechnology-related patent applications published for the first time, organized by country of the assignee, for different 4-year periods.

Another level of analysis examines whether patents are filed internationally versus only in the inventor's home country. This may be viewed as a measure of international impact, although there are other reasons why patents may not be filed internationally, such as the cost of patent prosecution abroad, or when the main market for the technology is in the home country. Of all nanotechnology patents filed internationally (the criteria being in three or more countries) during the 1985-2007 period, about 41 percent have U.S. assignees, while only 1.3 percent have Chinese assignees, as shown in Figure 3-5.

Based on these metrics there are two clear trends in China: Chinese inventors are filing patents at a rapidly increasing rate, and most of these filings are aimed at protecting innovations in their home market. China's patent trends, combined with its expected future economic growth, could afford it widespread nanotechnology protection in one of the world's largest markets in coming decades.

Dedicated nanotechnology funding

The United States still invests more money in nanotechnology R&D that any other country, with a total of $5.7 billion in investments in 2008, including $1.9 billion from Federal and State governments, $2.7 billion in corporate R&D, and $1.0 billion in venture capital invest¬ments. However, as in other areas, the rest of the world is closing the gap, and even surpassing the United States by some metrics. Asia as a region now invests more than the United States, totaling $6.6 billion in 2008. Of the Asian total, Japan leads with $4.7 billion in R&D funding, but with growing contributions from China, South Korea, and other countries, as shown in Figure 3-6. While the U.S. Government funds more R&D than any other individual country, Figure 3-7 shows that total nanotechnology R&D spend¬ing by the United States was eclipsed by Europe in 2005 and by Asia in 2007. From 2003 to 2008, total nanotech funding in the United States grew at a compound annual growth rate of 18 percent, while funding in the rest of the world grew at 27 percent annually.

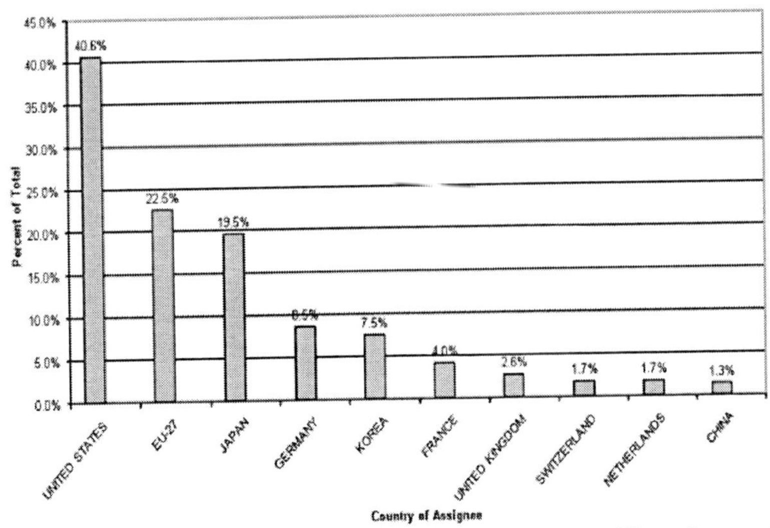

Source: Bruce Kisliuk, USTPO, unpublished study on comparative patent filings, January 2010

Figure 3-5. Inventions filed in at least three countries, organized by country of the assignee, cumulatively over the years 1985–2007.

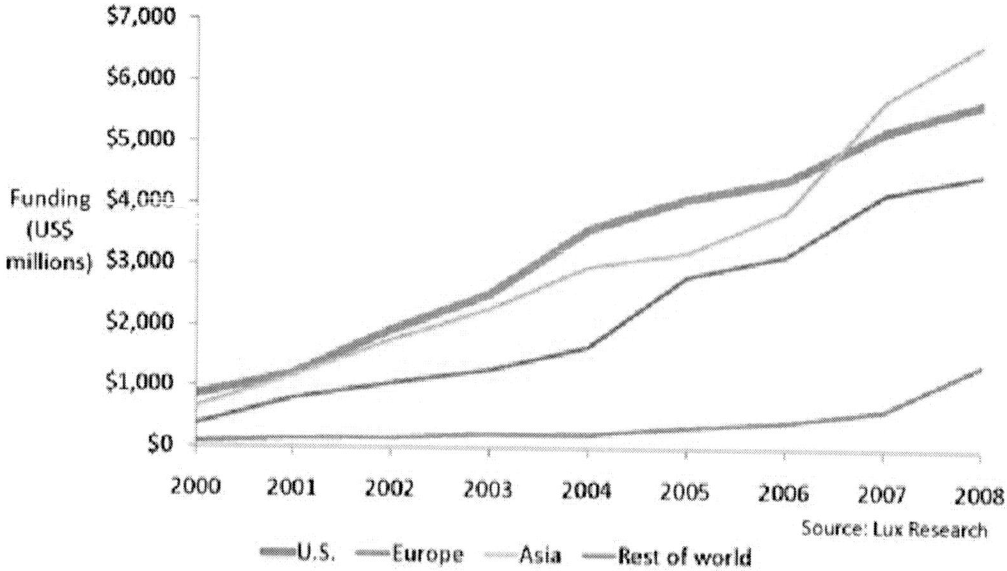

Source: Lux Research

Figure 3-6. Total funding for nanotechnology (from all sources, including government, corporate R&D, and venture capital), plotted by year, shows Asia in the lead since 2006.

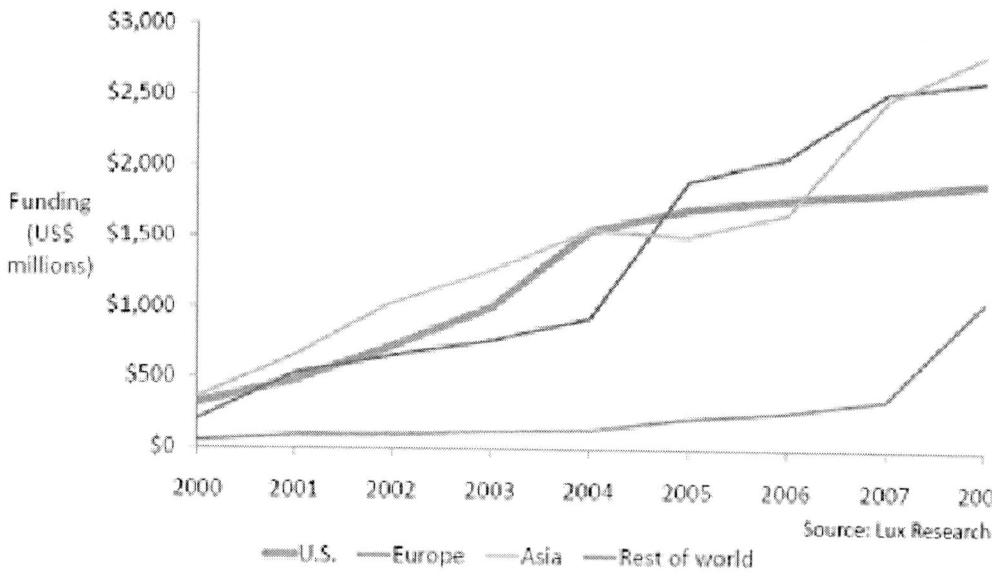

Source: Lux Research

Figure 3-7. Over the same period, government funding in the United States has lagged that in Europe and Asia.

Education and workforce development

Over the course of the past decade, the United States has educated more of the world's science and engineering workforce than any other country, largely through research and development funds provided through the NNI. However, the current rate of retention of highly trained non-U.S. citizens hampers the Nation's ability to capitalize fully on the opportunities created by nanoscience and nanotechnology. Consider the following statistics: From 2004–2007, 26,035 Ph.D. degrees were granted in science and engineering by U.S. universities to citizens from China, South Korea, and India. This figure represents 55 percent of all nonresident Ph.D. degrees, and 18 percent of total science and engineering Ph.D.s in that time period. Of these students, 87 percent indicated intent to stay in the United States after completing their degrees. Ultimately, however, only 57 percent of those indicating such intent remained in the United States. Thus, it can be inferred that on the order of 9,700 of these highly trained individuals were lost to the U.S. workforce during this time. The N NAP believes that the numbers for nanotechnology mirror these for science and engineering as a whole. U.S. national laboratories and companies that must hire U.S. citizens to conduct R&D in certain nanotechnology areas are especially handicapped by a severe shortage of qualified personnel. Relatively few students from foreign countries are naturalized while in graduate school, so the percentage of non-U.S. citizens graduating is probably the same as those entering.

Overall nanotechnology growth

Taking into account the above metrics, as well as the technology development capabilities of nations based on additional factors such as the percentage of gross domestic product (GDP) derived from high technology, R&D spending as a percentage of GDP, and trained workforce development (in addition to education and retention of science and engineering Ph.D.s), the United States remains the leader in nanotechnology R&D. However, Japan, Germany, and South Korea are closing the gap.[14] As of 2008, Japan ranks second with consistent strong investment in corporate R&D and numerous government initiatives. There has been explosive growth of nanotechnology R&D in Germany over the past several years as a result of its own investments combined with significant funding from the European Commission Seventh Framework Programme. In fact, German nanotechnology R&D has now surpassed that of the rest of the European Union. South Korea ranks third in corporate spending on nanotechnology and is perhaps better positioned to capitalize on nanotechnology than any other Asian nation, despite its small relative size. China is poised to capture manufacturing activity in value-added nanomaterials as applications grow and commoditization ensues. It is likely to develop a strong position in the low-cost manufacture of nanomaterials such as carbon nanotubes, nanofibers, and nanopowders. However, China is also improving its capabilities in areas not reliant solely on low- cost manufacturing by ramping up its homegrown scientific research and workforce talent pool. The inescapable conclusion is that the United States cannot assume that it will continue to be the world nanotechnology leader.

Technology Transfer

Nanotechnology is expected to have substantial and long-lasting impact in many areas, including nanoelectronics, healthcare, and clean energy. The electronics industry already manufactures products that involve nanoscience and nanoengineering innovations.

Applications in health care (diagnostics, delivery, and discovery) and clean energy technologies (efficiency, storage, and photovoltaics) are developing rapidly.

All told, U.S. corporations have invested an estimated $2.75 billion in nanotechnology R&D, 50 percent of which was spent by the electronics & information technology sector, 37 percent by the materials and manufacturing sector, 8 percent by the healthcare and life sciences sector, and 4 percent in the energy and environment sector.[15] Most large U.S. corporations in these sectors have established structured nanotechnology efforts over the last decade, though their approaches vary widely from loosely coordinated research initiatives to centralized task forces for mapping nanotechnology's impact.[16] Over the last two years, however, as companies have better understood the potential effect of nanotechnology on their business, their focus has shifted from evaluating nanotechnology as a broad theme to capitalizing on specific nanotechnology opportunities.[17] While large corporations are not often direct recipients of N NI funding, they will often serve as channels to market for nanotechnology-based products. In addition, these large companies rely heavily as sources of nanotechnology innovation and partnerships on the start ups, universities, and national laboratories that do benefit directly from NNI support.[18]

Many nanotechnology innovations are also being developed within start-up companies funded by venture capital (VC). Over the past 10 years, VC investment in U.S. companies developing nanotechnology totals $5.03 billion, accounting for 86 percent of the worldwide VC investments in nanotechnology. VC funding increased dramatically in 2008, reaching $1.15 billion, its highest total ever. However, as a result of the recession and mirroring the pattern in other technology areas, VC funding fell to $667 million in 2009. Over the last two years, VC involvement in nanotechnology was driven increasingly by large investments in later stage rounds for existing companies, as opposed to investments in new firms. The total number of new companies developing nanotechnology in the United States receiving seed or Series A financing in 2008 and 2009 totaled 56, down from 74 in 2006 and 2007. While some of the drop-off reflects the state of the world economy, venture capitalists are increasingly averse to areas of nanotechnology that have long times to market and high capital requirements, thus becoming more selective in their investments.[19] As a result, there is a need for novel approaches and funding mechanisms to support the transfer of technologies with long incubation times from the laboratory to the market.

Some nanotechnology start-up companies have begun to reach liquidity events, initial public offerings (IPOs) or acquisitions that provide returns to their investors. Despite some notable successes, returns to investors from nanotechnology start ups have been mixed. There is still a need for assistance to start ups, not only in bridging the gap from academic research to initial commercial funding (the traditional "valley of death"), but also in transitioning from venture-backed research and product development into the commercial production. It is difficult for venture capitalists and other financiers to earn returns on their investments until their companies begin generating significant revenues and, ideally, profits.

Over the past decade, seven U.S. nanotechnology firms have completed IPOs, with a cumulative value at the time of their IPOs of $3.75 billion. Since the last NNAP review, A1 23 Systems, a nanotechnology-based battery company, completed a $428 million offering, making it one of the most successful technology IPOs in 2009; the firm is considered a bellwether for the field of energy and environmental technologies.

Mergers and acquisitions (M&A) of nanotech firms have been more common but less lucrative. While 15 nanotechnology start-up companies based in the United States were acquired during the last decade, including six since the last NNAP review, many companies have been sold at a loss. The purchase price for those 15 acquisitions is estimated to total approximately $260 million, compared to $273 million that had been invested into those companies prior to their acquisition. Finally, many nanotechnology start ups have simply gone out of business, including at least 15 VC- backed firms.

In the health sciences, nanotechnology has evolved from early state demonstrations and discovery to a more mature phase in which a portfolio of practical applications is emerging. Here, progress in technology transfer can be assessed by the nanotechnologies expected to have substantial clinical impact. Two areas are particularly relevant:

- **Diagnostics and imaging**—Nanosensors are detecting clinical biomarkers with higher sensitivity and specificity as compared to existing assays. Sensitivities for proteins and nucleic acid detection are quickly moving into the femtomolar and attomolar range, a level previously unobtainable and one that is creating new opportunities for advancing the detection of disease. Imaging is being enabled by new contrast agents that are active in a wide range of clinical modalities, including optical imaging (quantum dots), magnetic resonance imaging (iron oxide particles), and ultrasound (polymeric bubbles). These nanoscale imaging contrast agents are enabling clinicians to view organs at a level of detail that has not been easily obtained until now. Finally, nanostructures such as carbon nanotube electron emitters are being employed to build a new generation of imaging instrumentation that have lower voltage demands, multiplexed emissions sources, and smaller footprints. Some of these new instruments are likely to reduce patient radiation exposure while producing images with enhanced resolution and sensitivity.
- **Therapeutics**—A wide range of nanotechnologies is effectively enabling nearly all phases of drug development. Notably, drug delivery nanotechnology platforms have advanced significantly in the last five years, with increasing capability to deliver drugs to specific organs and cell types. Progress in moving new nanotechnology-based drugs out of the laboratory is significant with a number of clinical trials and Investigational New Drug applications (INDs) filed from investigators funded by the National Institutes of Health. These nanotechnologies are also being placed on a commercialization path, as demonstrated by the numbers of new companies emerging from various areas of nanotechnology supported through NIH. To accelerate the translation of research into clinical advances, the National Cancer Institute and FDA are working together to develop a clear pathway for evaluating and approving a diversity nanotechnologybased diagnostics and therapeutics. Strengthening such interagency coordination is essential for successfully moving nanotechnology from the laboratory into the commercial arena.

Barriers to Commercialization

The N NAP observes that the rate of innovation and the rate of opportunity creation have not decreased over the first 10 years of the NNI, including during the two years since the last NNAP review. However, against the backdrop of immense opportunity, there is also a clear sense, based on metrics such as those discussed earlier, that the United States is losing ground to other countries in the scale up, commercialization, and industrialization of new nanotechnologies. The NNAP believes that there are a number of significant barriers that hinder the ability of the United States to most effectively capture the benefits of the Nation's nanotechnology research endeavors, which the NNI has worked so hard to promote. The NNAP also believes that there are concrete steps that the NNI can take to reduce or eliminate these barriers.

RECOMMENDATION 3-1. NANOMANUFACTURING AND COMMERCIALIZATION

The NSF, DOE, Department of DOD, NIST, and NIH should include a greater emphasis on manufacturing, and commercialization while maintaining or expanding the level of basic research funding in nanotechnology. Specifically, over the next five years, the Federal Government should double the funding devoted to nanomanufacturing (PCA5). In addition, the Federal Government should launch at least five government-industry-university partnerships, using the Nanoelectronics Research Initiative as a model. The Federal Government should also support at least five Signature Initiatives over the next two to three years, with each Signature Initiative funded at levels adequate to achieve its stated goals, presumably between $20 million and $40 million annually.

Successful commercialization of any developing technology requires a healthy ecosystem in which there is continuous innovation at all levels; it cannot simply be based on exploitation of past accomplishments. Fundamental research provides the seed corn for innovation in nanotechnology, without which the pipeline quickly empties. Thus, support for basic research in nanotechnology must not be reduced.

Traditional academic boundaries impede crosscutting research, as do funding silos where particular topical areas are captive to, or perceived to be captive to, single agencies. Since nanotechnology science and engineering is truly interdisciplinary, NNI member agencies should remove obstacles that prevent them from collaborating more. As part of that effort, there should be more interaction among Federal Government funding agencies to remove barriers to multi-agency support during the lifecycle of R&D projects. As an example, organizations that have successfully completed a Phase II NSF grant may next benefit from bridge funds offered by other government agencies, e.g. NIH; such transitions should be free of obstacles.

Moreover, as nanotechnology matures and concepts reach practical application, a greater emphasis on research in nanomanufacturing and commercial deployment is appropriate; nanomanufacturing refers to efficient and scalable methods for producing nanoscale materials and devices. Research in this area also includes development of advanced nanofabrication tools, new nanomaterials characterization methods with high throughput and differentiation, nanomaterials purification technologies, and nano- materials standards. In addition, the NNI can help promote successful commercialization by supporting applied and translational research on the integration of nanoscale materials and devices into useful products—incorporating carbon nanotubes into composite materials or optimizing biohazard sensors for performance in battlefield environments, for example. Moreover, the establishment of standards is essential to growth of most new technologies, and nanotechnology is no exception. One approach to facilitating standards development would be to establish a

"Particle Foundry" that produces standard particles of common nanomaterials of controlled size, shape, and composition for scientific researchers and industry.

Advances in all of these areas are necessary for nanotechnology to realize its enormous technological and societal impact. Nanomanufacturing and commercial deployment should also be a key component of a broader national mission in advanced manufacturing innovation that would allow the United States to level the playing field in competition with countries with labor cost or natural resources advantages. As part of this emphasis, the NNI should continue to establish new Signature Initiatives. One of these could be the development of the "Desktop Fab," that is, a system that has the ability to make prototype versions of nanostructures rapidly at the point-of-use, an effort that would require a large multi-million dollar facility today. Such a development could represent a manufacturing advance with impact analogous to that of desktop printing in personal computing. Other challenges could include setting goals such as creating low-cost solar cells or building lightweight ultra-efficient vehicles, to spur applied and translational research directed toward those applications.

RECOMMENDATION 3-2. JOB CREATION

The Department of Commerce and the Small Business Administration should advise the NNI on how to ensure that its programs create new jobs in the United States, including coordinating with State efforts, and economic impact should be an explicit metric in the second decade of the NNI.

During its second decade, the NNI should adopt an explicit goal to focus research in areas that by the end of the decade results in measurable job creation in the United States. Taxpayers deserve a return from the public research dollars invested in nanotechnology, and the NNI should account for the impact this research is having on the economic conditions of average Americans, and direct future research to areas that are likely to benefit American firms. The NNI should have some funding of its own that it can direct toward encouraging entrepreneurship in the nanotechnology field. The NNI could use such funding to help companies more successfully navigate the "valley of death" and to develop manufacturing capability in the United States for producing strategic nanotechnology products.

As noted earlier, the United States educates more of the world's talent in nanoscience and engineering than any other country, and then exports much of that talent. One solution would be to develop a program to provide U.S. Permanent Resident Cards for foreign individuals who receive an advanced degree in science or engineering at an accredited institution in the United States and for whom proof

RECOMMENDATION 3-3. WORKFORCE RETENTION

Congress and the Administration need to take steps to retain scientific and engineering talent trained in the United States by developing a program to provide U.S. Permanent Resident Cards for foreign individuals who receive an advanced degree in science or engineering at an accredited institution in the United States and for whom proof of permanent employment in that scientific or engineering discipline exists.

Innovative technologies cannot be developed on a series of $100,000, nine-month grants when the fully-burdened cost of supporting a single Ph.D. scientist in a research company is about $300,000 annually. Therefore, the N NAP believes that the N NI needs to make sustained and substantial investments in focused areas. Among existing government programs, the NIST Technology Innovation Program (TIP) is a good model; it provides for multi-year, multi-million-dollar investments to help companies developing new nanotechnologies cross the infamous "valley of death." VC-backed start ups should not be excluded from Federal Government grants to facilitate commercialization.

Recommendation 3-4. Moving Nanotechnology to Market

The DOE, DOD, NIST, NIH, NCI, FDA, and NIST should clarify the development pathway and increase their emphasis on transitioning nanotechnology to commercialization, including making sustained meaningful investments in focused areas to help accelerate technology transfer to the marketplace.

Case Studies

During the past decade, the nanotechnology R&D community has made an enormous number of discoveries with the potential for wide-ranging societal impacts. The majority of these developments remains at embryonic or proof-of-concept stages and have not yet begun a trajectory toward the marketplace. Among the great many significant discoveries that could become the basis of future commercial technologies are:

- **Graphene transistors**—The use of graphene (single layers of carbon atoms in a chicken wire geometry) as the semiconductor in transistors promises to improve performance and keep the electronics industry on the path of miniaturization, but researchers need to learn to control the material's properties before they can realize its potential. With strong support from the semiconductor industry, commercialization could come within the next decade.
- **Nanomotors**—These molecular machines, akin to a cell's enzymes, have the potential to revolutionize the way researchers control atoms and construct materials. However, investigators pursuing nanomotors are breaking into uncharted territory; valuable and practical implementations remain decades away.
- **Metamaterials**—These synthetic materials enable the manipulation of electromagnetic (EM) waves in new ways, offering the potential for significant advances in communication technologies and the control of light. Early simple applications may reach the market in 10 years.
- **Silicon nanowire thermoelectrics**—Silicon nanowires could significantly lower the cost of thermoelectric devices that convert heat directly into electricity, paving the way for broader adoption. Widespread waste-heat recovery would have a worldwide, immediate impact. Another 10 years of development is likely necessary before silicon nanowire thermoelectrics will be ready for the spotlight.

- **Plasmon-enhanced solar cells**—Properly designed metal nanoparticles can scatter incoming light, and when laid down as a layer can greatly increase efficiencies of solar cells. Practical demonstrations have been made and scalable processes for making the nanoparticles are available, opening a potential pathway to bringing plasmon-enhanced solar cells to the market in about five years.

Some nanotechnologies have matured beyond early stages of development and represent significant progress toward the goal of delivering commercially consequential technology that NNI has fostered directly and indirectly. The case studies that follow specifically focus on commercialization milestones since the last NNI review.

Advanced Batteries—The small size, light weight, and high-energy density of lithium ion batteries, first commercialized in the early 1 990s, proved perfect for consumer electronics devices such as laptop computers and cell phones. However, these batteries were not suitable for large-scale energy applications. To meet the needs of large-scale applications, A1 23 Systems (Hopkinton, MA) developed its patented Nanophosphate™ technology, based on nanoscale lithium metal phosphate cathodes with compositionally and structurally engineered phase stability and lithium transport kinetics. A1 23's technology provides the necessary combination of high energy density, high power, safety, long life, environmental friendliness, and low cost to enable applications in transportation and the electric grid.

The company, founded in 2001 to commercialize technology developed at MIT with DOE funding, brought its first product to market in 2005, a cell developed to power a new line of premium, contractor-grade cordless power tools for Black and Decker's DeWalt product line. Today, the company has more than 2,000 employees and is at the forefront of creating a battery industry in the United States that will help change the way the Nation produces, stores, and uses energy. A1 23 has built on its core technology to develop advanced batteries and entire battery systems to help bring in a new era of sustainable transportation, including plug-in hybrids and battery electric vehicles for both passenger and commercial vehicles.

In addition, A1 23 is developing new energy storage solutions for the electric grid across the entire range of applications including generation, transmission, and distribution. These products will help create a power generation and delivery system that is more efficient and makes the most of existing power-generating assets. These products also will assist in the integration of renewable energy sources, such as wind and solar power, by smoothing their output and stabilizing the grid, thus allowing those technologies to achieve their full potential. Development of these nanotechnologybased products will reduce reliance on foreign oil, invigorate advanced manufacturing in America, and help meet the global imperative for greater energy efficiency.

The company was started using an initial $100,000 SBIR grant from the DOE. Since then, A1 23 has raised $350 million in private capital from investors that include GE, and was awarded a $249 million DOE battery manufacturing grant in 2009. The company is expanding its lithium ion battery manufacturing facilities in the United States, beginning with facilities in Michigan that will stimulate the economy through the creation of high quality jobs in the short term, and play a key role in building a stable, self-sustainable and globally competitive U.S. battery industry in the long term. The company raised $428 million in an IPO in September of 2009 that was one of the largest of the year.

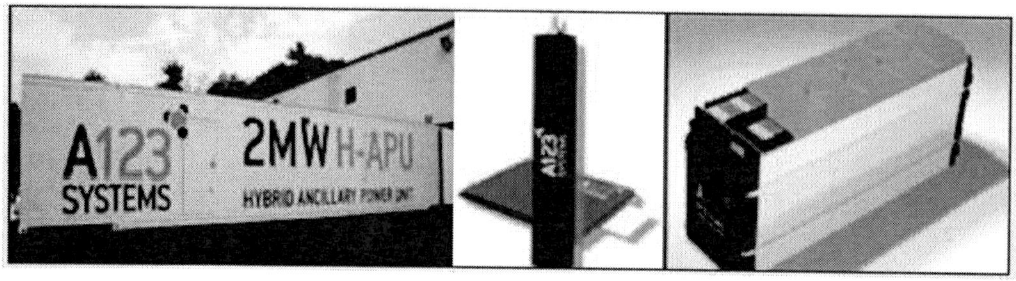

Lightweight Wiring and Body Armor—Today's military conducts complex missions against unpredictable threats. A key to mission success is the use of highly digitized, network-centric and space-based systems to provide command and control, reconnaissance, and communications for the warfighter and national command authority. Ironically, these state of the art systems are dependent upon 1 9th century technology—heavy copper wires and cables. One third of the weight of a satellite is found in wiring harnesses, and given that it costs up to $100,000 to lift one pound of mass into orbit, the use of copper comes with a significant weight penalty. Similarly, a civilian airliner such as the Boeing 787 has over 60 miles of copper wire, weighing many thousands of pounds and thus providing a serious drag on fuel economy.

A 21st century replacement for copper is at hand, thanks to NNI-associated activities. Nanocomp Technologies, Inc., (Concord, NH) has spent the last five years developing a process technology that grows ultra-long carbon nanotubes (CNTs) and simultaneously fabricates them using automated equipment into sheets and spun yarns. Working with Northrop Grumman's Aerospace Division and the Air Force, Nanocomp has developed a CNT-based electrical core conductor and cable shielding that can be used as a lightweight, robust substitute for shielded copper wiring. For the first time, CNT cables have achieved high fidelity USB 2 data transmission rates equal to that of off-the-shelf cables. In 2009, the company demonstrated that Category 5 CNT materials are reducing the weight of cable harnesses by 33–70 percent, with the potential of eliminating hundreds of pounds per spacecraft and thousands of pounds per aircraft. Weight reduction of this scale could produce fuel savings totaling tens of millions of gallons over the lifecycle of a commercial jetliner and reduce the environmental impact of commercial air travel by reducing carbon dioxide emissions at a scale of hundreds of millions of pounds.

Additionally, work funded by the Air Force and Lockheed Martin has shown that Nanocomp's nanotube sheets significantly improve the resistance of military aircraft and satellites to electromagnetic interference and electromagnetic pulses. NASA's upcoming Juno space mission to Jupiter will, for the first time, use nanotube sheets to protect the spacecraft from electrostatic discharge. Civilian aircraft will also benefit as these same CNT materials can enhance the protection of avionics systems. Ultimately CNT materials developed for military shielding will find use in consumer electronics, such as smart-phones, that also require high performance EMI shielding to assure their operation.

Since 2005, the Army's Natick Soldier Center has supported work to use the company's CNT materials to help decrease the weight and improve protection of body armor. In 2009, the company's nanotube composites successfully defeated bullets – a promising milestone for lightweight body armor. CNT based lightweight armor will also find broad use in ground vehicles, helicopters, and fixed wing aircraft. The armor program has also yielded strong, lightweight composite materials for military satellites and aircraft with the potential to address energy savings in civilian markets such as windmill blades and automotive products.

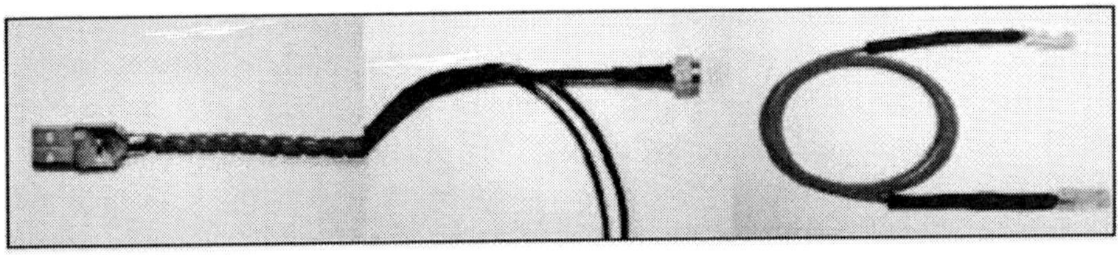

Fully Functional CNT cables from left: USB, Coaxial, and Cat 5 Internet cables.

Antimicrobial Technology—The intense focus on nanotechnology created by the NNI over its first decade has encouraged many established companies to start internal nanotechnology research and development efforts. For example, Baxter International (Deerfield, IL) has developed and commercialized the Vitalshield® technology that uses silver nanoparticles to reduce catheter-related blood stream infections (CR-BSIs) and serves as an example of the leverage that the NNI provides in product and job creation even when direct funding is not provided.

It is estimated that more than 400,000 cases of CR-BSI occur each year in the United States. CR-BSIs increase average duration of hospital stay by 23 days and raise mortality by 18 percent. The increase in direct costs to hospitals has been estimated to range from \$13.8 billion and \$22 billion annually. While migration of skin organisms at the catheter insertion site is a common route of infection, contamination of the catheter hub contributes substantially to intraluminal colonization of longterm catheters.

In Vitalshield®, the silver nanoparticles create a permanent, nonflaking coating on catheters that continually generates a thin layer of antimicrobial silver ions. Each nanoparticle in the structure develops a surface layer of silver oxide, which serves as a source of ionic silver upon contact with moisture. This oxide layer releases ionic silver into the surrounding fluid space and re-develops as a new layer of elemental (metallic) silver is exposed. This process continues with ongoing exposure to moisture providing a "reservoir" of ionic silver that is durable and persistent. Data show that the silver nanoparticles in Vitalshield® elute more than three times the available silver ions over a 96-hour period than zirconium phosphate embedded silver technology or in silicone elastomer.

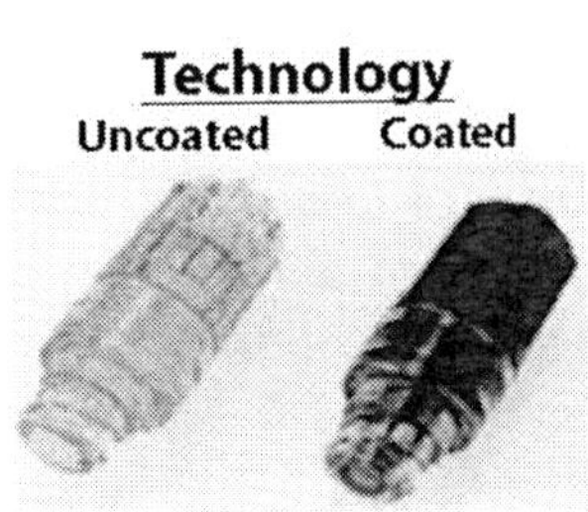

IV. Nanotechnology and Environment, Health, and Safety Issues

Chapter Summary

By creating jobs, stimulating economic growth, and providing solutions to some of the toughest challenges facing humankind, nanotechnology has great potential to change the world for the better. Yet realizing this potential may be thwarted if the safety of new materials and products arising from nanotechnology is not addressed up front. In the absence of sound science on the safe use of nanomaterials and of technologies and products containing them, the chances of unintentionally harming people and the environment increases. At the same time, uncertainty and speculation about potential risks threaten to undermine consumer and business confidence. The proactive approach to addressing potential environment, health, and safety impacts of nanotechnology taken by NNI is commendable. Over the past two years, the NNI has released a cross-agency nanotechnology EHS research strategy, instigated multi-stakeholder workshops on nanotechnology EHS issues, and seen the Federal nanotechnology EHS research budget increase from $67.9 million in 2008 to a requested $1 16.9 million in 2011. Individual agencies have also played an active role in international efforts to develop nanotechnology responsibly.

Even so, potential EHS-related barriers still stand in the way of effective, sustainable, and responsible commercialization of nanotechnology. As the NNI continues to work toward "a future in which the ability to understand and control matter at the nanoscale leads to a revolution in technology and industry that benefits society," the NNAP recommends that member agencies increase coordinated efforts to overcome these barriers. Specifically, the NSET Subcommittee's interagency working group on Nanotechnology, Environmental, and Health Implications should develop clear principles to support the identification of plausible risks associated with the products of nanotechnology. The NSET Subcommittee's NEHI working group should also further develop and implement a cross-agency strategic plan that links EHS research activities with knowledge gaps and decision-making needs within government and industry. The NNAP also recommends that NSET Subcommittee implement organizational changes that support consequential cross- agency action on addressing nanotechnology EHS issues. In particular, the NNCO should specify an individual who would lead interagency coordination of efforts in the area of EHS. Finally, the NSET Subcommittee's NEHI working group should develop publicly-available information resources on cross-cutting nanotechnology EHS issues that are relevant to businesses, health and safety professionals, researchers and consumers.

Manufacturing, known in some sectors as fabrication, is a significant wealth-producing component of the American economy, and because manufacturing may pose a number of risks, manufacturing activi¬ties are governed by Federal labor laws and environmental laws. Whether manufactured nanomaterials pose unique risks is an important question. As nanotechnology has developed over the past decade, those investing in, developing, using, or regulating products and processes have faced a significant challenge: how to ensure safe and responsible use where potential risks are uncertain. In recent years, media coverage has highlighted uncertainties over potential nanotechnology-related risks; experts have called into question the ability of governments and industry to ensure the safety of nanotechnologybased products; scientists have identified potential risks associated with a number of nanomaterials; and Congress has emphasized the need for the NNI to support the responsible development of nanotechnology. To some, the result has been an emphasis on EHS issues that threaten to raise unsubstantiated concerns, undermine consumer and investor confidence, and create regulatory barriers to innovation. These are valid concerns. And yet, rather than calling into question the relevance of EHS research, they emphasize the dangers of *not* addressing environment, health, and safety issues upfront.

Responsible and sustainable development depends on a commitment to develop and use products that meet human and societal needs, while making every reasonable effort to anticipate and mitigate adverse effects and unintended consequences. In the context of nanotechnology, responsible and sustainable development relies on science-informed and socially-responsive approaches to identifying, assessing, and managing risks. Research to date suggests that some products of nanotechnology have the potential to present new or unusual risks to human health and the environment. For instance, nanoscale particles may penetrate to places in the body that are inaccessible to larger particles; radical changes in behavior at the nanoscale may render harmful materials considered to be safe in larger-scale and more conventional forms.

In general, nanomaterials that have been shown to pose risk fall in two categories. They are either derived from bulk materials that are known to pose EHS risk (e.g. heavy metals), or they are nanomaterials made from bulk materials that are generally considered nontoxic but, when miniaturized and aerosolized, exhibit increased risk.

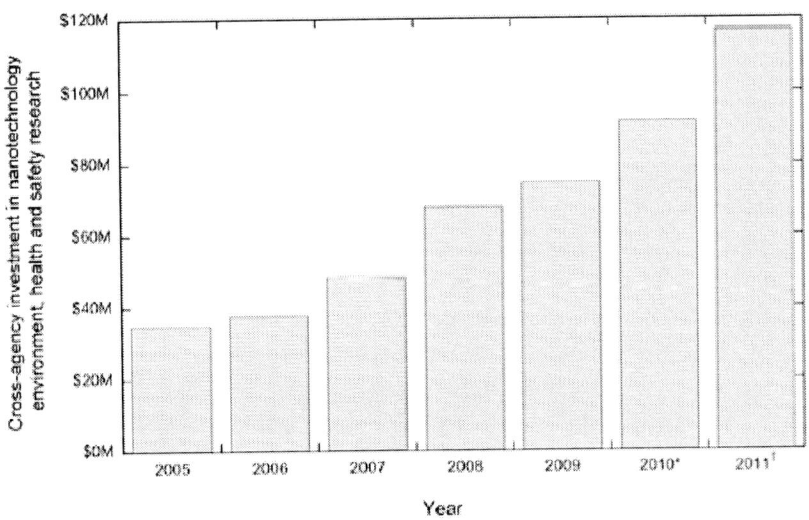

Source: NNI supplements to the President's budget, 2007–2011.

Figure 4-1. Cross-agency investment in nanotechnology EHS research, 2005–2011.

*Estimated. †Proposed.

In the absence of more detailed scientific evidence—and effective assessment and communication of the evidence that does exist—the distinction between plausible and implausible risks remains unclear. The resulting uncertainty threatens to undermine confidence and trust amongst investors, businesses, and consumers, and could jeopardize the success of nanotechnology. This is not a hypothetical threat. Consumer and advocacy groups already have raised concerns over the use of engineered nanomaterials in products as diverse as clothing, fuel additives, and sunscreens. Businesses have been hampered by regulatory uncertainty. A number of industries have shied away from nanotechnology for fear of consumer rejection in the face of speculative concerns.

Against this backdrop, it is no surprise that the potential environment, health, and safety impacts of nanotechnology have been a central focus of the NNI for the best part of the past 10 years.

Progress on Addressing the 2008 NNAP Review

In proactively addressing nanotechnology EHS issues, the collective efforts of the NNI member agencies have broken new ground. A commitment to the responsible development of nanotechnology is evident both within the NNCO and the NNI member agencies. Significant progress has been made in funding relevant research since the inception of the NN I and in the two years since the previous NNAP review,[20] as shown in Figure 4-1. Over the past two years, for example, NSF and EPA have jointly funded two Centers on the Environmental Impacts of Nanotechnology; EPA has collaborated with the British Environment Agency to fund nanotechnology EHS research; and the National Institute for Occupational Health and Safety (NIOSH) has increased intramural research efforts addressing potential health impacts from engineered nanomaterials in the workplace.[21] In addition, individual agencies are beginning to develop responsive internal strategies to specific nanotechnology EHS challenges. The number of publications on nano-related EHS issues is rising (see Figure 4-2), and the NNI EHS strategy[22] continues to mature with a program of ongoing, multi-stakeholder workshops feeding into its next iteration.

Amidst all of this progress though, a question remains: Is enough being done to identify and mitigate potential risks and reduce EHS-related barriers to effective, sustainable, and responsible commercialization of nanotechnology-based products?

In 2008, the N NAP acknowledged the NN I's role in addressing nanotechnology EHS issues and supported the soundness of the NNI's approach to addressing potential risks arising from nanotechnology. Through 10 recommendations in its 2008 report (restated in bold below), NNAP pointed the way to further strengthening the NNI's activities to ensure the responsible development of nanotechnology. The NNAP believes that in the two years since then, the NNI has made progress in addressing these recommendations, many of which remain relevant in 2010. In this rapidly developing field, a number of new priorities also have emerged that will shape the future perspectives and activities of the N NI. In what follows, the NNAP focuses on the progress the NNI has made regarding NNAP's 2008 recommendations and consider how the NNI might continue to support the emergence of responsible nanotechnology in the future.

1. ***Coordinate nanotechnology ENS strategy with industry and international stakeholders.*** Over the past two years, the agencies comprising the NNI have made substantial progress in coordinating the nanotechnology EHS strategy with industry and international stakeholders. Much of this progress has been led by individual agencies. The EPA, FDA, N IOSH, and the National Institute of Environmental Health Sciences (NIEHS), for example, have increased their level of interaction with industry and other stakeholders. Moreover, U.S. representatives continue to play key roles in nanotechnology EHS initiatives within the OECD, the International Organization for Standardization (ISO), and other international organizations and initiatives. Within the NNI, a series of workshops have been held that enable broader stakeholder input to developing and implementing a strategy for EHS research in the Federal nanotechnology portfolio. Although the NNAP feels that these workshops have occurred rather late in the process and do not fully address the degree of coordination required with industry and other stakeholders, they are an important and positive step forward. Nevertheless, there remains a need for greater coordination with industry and with other stakeholders that leverages nongovernment expertise and initiatives while minimizing unnecessary duplication of effort. Initiatives supporting further national and global partnerships between government, industry, and academia in addressing common nanotechnology EHS challenges are encouraged. Specifically, the NNAP suggests that channels of communication between relevant State and Federal agencies be established, which support information sharing on emerging nanotechnology EHS challenges and their potential solutions. Regarding industry, the sharing of nonproprietary information between industry and government remains difficult under existing frameworks and approaches to overcoming this barrier to progress should be investigated.
2. ***Do not segregate implications research and applications research. In many instances, nano- technology ENS research cannot be separated from the particular application(s) research and from the context for which a specific nanomaterial is intended.*** In contrast with the previous NNAP review, the members of the current NNAP believe that EHS research must be closely linked to applications research but that too great an emphasis on integration has the potential to jeopardize both applications and EHS research. Overly intimate integration of these two categories of research can lead to a muddling of research priorities. At the same time, there is a danger of resource-intensive EHS research compromising exploratory and applications-based research programs. To avoid those outcomes, the NNAP encourages a collaborative approach among communities involved in exploratory research, applications research, and EHS research that allows risk-researchers the freedom to identify and address issues from the perspective of protecting human health and the environment. To help achieve this, the NNI should encourage greater communication between communities developing nanotechnology-based applications and those addressing potential risks, and should foster a culture of mutually beneficial interdisciplinary collaboration between these communities. Importantly, the NNAP emphasizes the need to establish EHS research projects and programs that are led by researchers with expertise in identifying and addressing potential risks.

3. ***Continue developing joint programs among NNI agencies that leverage expertise and resources to conduct nan otechnology EHS research and to support agency missions.*** Member agencies of the NNI continue to work together and to leverage internal resources in addressing nanotechnology EHS. Over the past two years, EPA and NSF have jointly funded two major centers addressing environmental implications of nanotechnology. These investments will total some $38 million over the next five years. NIEHS, NIOSH, and FDA continue to collaborate on supporting research into the safety of nanomaterials through the National Toxicology Program. NIOSH and EPA have also worked together on developing workplace safety requirements for nanomaterials under the Toxic Substances Control Act. The NNAP encourages NNI member agencies to continue developing joint programs. In particular, the NNAP encourages stronger collaborations between oversight-focused agencies, which lack funding resources for research, and agencies with substantial research resources. Specifically, the NNAP encourages the development of risk-relevant research programs supported by research agencies such as NSF and NIH that are responsive to the perspectives and decision-making needs of regulatory agencies such as FDA, EPA, CPSC, and the Occupational Safety and Health Administration (OSHA).
4. ***Support wide distribution and availability of new nonproprietary information about the properties of nanomaterials.*** Following through on this recommendation remains important within the United States if nanotechnology-based products are to be developed responsibly and successfully. The NNAP is concerned that the recommendation, as it was stated in 2008, misses an important requirement: it does not address directly the challenge of ensuring open access to health and safety data while safeguarding confidential business information (CBI). Over the past two years, there has been only modest progress in collecting and disseminating non-proprietary information within the Federal Government. The recently ended EPA Nanoscale Materials Stewardship Program,[23] for instance, received a limited response from industry. The NNAP strongly encourages the NNI to rise to the challenge of developing mechanisms that allow broad access to EHS-relevant data by state and local regulatory and enforcement agencies, as well as the public, while respecting legitimate CBI claims. The NNAP also encourages the NNI to develop more effective mechanisms for rapidly disseminating information generated within the Federal Government and by State agencies. At the time of writing this N NAP review, for example, the results of an NNI workshop on exposure assessment held on February 24–25, 2009, have yet to be published.[24] Shortening the period between information gathering and information dissemination from years to months will support coordinated and strategic action on identifying and addressing nanotechnology EHS issues in a timely manner.
5. ***Assess the Federal nanotechnology EHS portfolio and update gap analysis against research priorities triennially.*** In February 2010, the NRC launched a four-year study, funded by the EPA, which will contribute significantly to addressing this recommendation by the prior N NAP. The statement of task for the study requires that the panel, "will create a conceptual framework for environmental, health and safety-related research; implement this research plan; and subsequently evaluate

research progress over a three year period of time." The first report from the project is scheduled to be completed by midyear 2011.

6. ***Leverage opportunities to bootstrap identified gap areas and to encourage increased investments elsewhere through collaboration with industry and other countries; encourage broad and ongoing agency participation in such efforts.*** Agencies within the NN I continue to leverage opportunities to collaborate with industry and other countries. Collaborations between Federal agencies, industry, and the OECD, ISO, and the United Kingdom Environment Agency continue to leverage international activities and investments. The NNAP strongly encourages that these activities be continued and expanded.
7. ***Encourage supported researchers to report on the development of analytical methodologies used in their research so that a series of best practices can evolve for risk assessment and characterization.*** The lack of good practices for materials characterization and approaches to risk analysis remains a critical challenge to conducting robust EHS-related studies and ensuring that risk-focused research is suitable for informing risk-based decisions. Over the past two years, Federal employees have been associated with initiatives addressing good practices in nanotechnology EHS research, including the MinChar Initiative,[25] the International Alliance for NanoEHS Harmonization,[26] and activities within the ISO and the OECD Working Party on Manufactured Nanomaterials. The NSET Subcommittee, however, has not taken a clear leadership role in supporting federally funded researchers in contributing to risk assessment and the development of good practices in materials characterization. The NNAP strongly recommends that the NSET Subcommittee take leadership roles in this area.
8. ***Promote broad and practical use of ENS findings in defining responsible use of nanotechnology in research, manufacturing, and commercial application.*** Although the NNAP feels that agencies within the NNI have made progress on this recommendation over the past two years, there is room for significantly more progress. In particular, where EHS issues cut across agencies, the NNI has a unique opportunity to act as a central point of contact for information and guidance, an opportunity that it has not yet taken. One approach to promoting broad and practical use of EHS findings would be to develop general principles for utilizing such findings. These principles would ensure alignment and consistency among the agencies with various oversight responsibilities as they propose actions based upon the EHS findings. The NNAP recommends that the NNI develop a set of such principles.
9. ***Increase exposure assessment funding.*** Exposure assessment research has been receiving more attention within the NNI over the past two years. NIOSH is collaborating with the CPSC on studying the potential of human exposure from use of selected spray applications utilizing nanomaterials. In February 2009, the NNI held a workshop on human and environmental exposure assessment that led to input from a range of stakeholders. In addition, the FY 2011 budget request for nanotechnology EHS research will provide increased funding for exposure assessment research. Nevertheless, a survey of the International Council on Nanotechnology (ICON) nanoEHS Virtual Journal[27] indicates that only 13 percent of the collected papers (spanning 2000–2009) directly address exposure, as shown in

Figure 4-2. The NNAP strongly recommends that funding for research on nanomaterial exposure issues be increased further, and that these increases are sustained over a number of years to ensure the development of robust and relevant exposure assessment tools.

10. ***Maintain and strengthen agency support and coordination efforts through the NSET Subcommittee and its NEHI working group.*** In early 2010, the director of NIOSH was established as co-chair of the NSET Subcommittee's NEHI working group. This is a welcome move, demonstrating senior-level agency support for the NSET Subcommittee's NEH I working group and the NSET Subcommittee. The NNAP encourages the NNI to follow this lead and work toward engaging other agencies at a senior level, ensuring that the NSET Subcommittee's N EHI working group and the NSET Subcommittee are in a position to contribute to high-level decision making within member agencies.

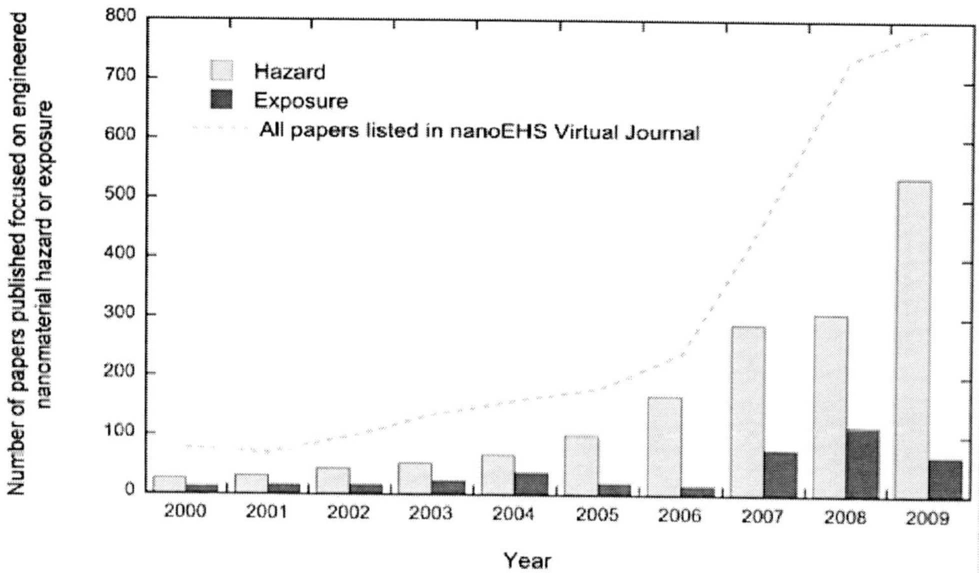

Figure 4-2. Number of published papers listed in the ICON nanoEHS Virtual Journal that focus on engineered nanomaterial hazard and exposure, compared to the overall number of papers related to nanotechnology EHS. Listed papers represent global publication output.

Hurdles to Future Progress in Addressing Nanotechnology EHS Issues

Looking to the future, the N NAP has identified six potential hurdles that N NI leadership faces as it strives to make further progress on ensuring the responsible development of nanotechnology:

1. ***Leadership and accountability***— Clear leadership and accountability continues to be important in identifying and addressing crosscutting issues and ensuring a sound risk-based approach to R&D and its subsequent use. Within the current NNI structure, the NNAP has not been able to identify an accountable individual who has

direct experience in risk research, understands what it takes to translate that research into decision making, and is in a position to engage with member agencies at a senior level. Similarly, it is not clear who, or which organization, is account¬able for cross-agency actions on ensuring the development of responsible nanotechnology, including implementing the recommendations in this review. Leadership and accountability on nanotechnology EHS is essential to raising confidence in the NNI within business and consumer communities and to enabling the United States to take a leadership role on the international stage.

2. ***Stakeholder engagement***—The NNI needs to more actively engage stakeholders on nano- technology EHS issues if it is to address these issues effectively. Engagement so far has involved participation in nongovernment and international initiatives, public hearings, and, more recently, in a series of EHS-related workshops. These efforts need to be expanded substantially if Federal research strategies are to reflect stakeholder needs and the current state of the art with nanotechnology EHS.
3. ***Connecting research to decision making***— To encourage the responsible and successful development and introduction of nanotechnology-based products, stakeholders need to work harder to bridge the gap between nanotechnology EHS R&D and decision making. The NNAP suggests that applied EHS research should be driven, at least in part, by the need to make informed decisions on risk assessment and management, and on materials and product oversight. The current structure of the NNI has not facilitated effective linkage of risk research and decision making, and this disconnect has resulted in actions and strategies that do not fully address policy needs. At the moment, in lieu of NNI coordination, individual agencies are unilaterally orchestrating effective connections between research and decision making. EPA and NIOSH, for example, have targeted their nano-related research strategies to fill knowledge gaps that are impeding decision making. Similar efforts are needed at an interagency level to ensure that cross-agency issues, such as dealing with uncertainty, risk assessment, and evidence-based decision making, are addressed appropriately.
4. ***Framing the ENS risk issue***—In framing nanotechnology EHS research within the context of an initiative focused primarily on promoting exploratory research and innovation, some science-based questions appropriate to EHS impacts are being missed. The NNI definition of nanotechnology has limited relevance to addressing new risks. Yet regulators and risk assessors are being pushed to address materials and products that fit inside this definition of nanotech¬nology, rather than being encouraged to develop a science-based approach to risks that is not bound by arbitrary (from a risk perspective) definitions. As a result, the focus is on size as a determinant of risk, rather than emerging risks *per se*. Continuing to frame EHS issues within the current terms of the NNI runs the double danger of raising illegitimate concerns that have no evidentiary backing, while obscuring real ones.
5. ***Research strategy***—The responsible development of nanotechnology is dependent on a clear multi-stakeholder research strategy that ensures that regulators, businesses, consumers, and others have the information and tools they need to develop and use new products safely. While the NNI has been active in developing a Federal strategy, the effort has some way to go before it fully addresses the needs of stakeholders. In 2009, an independent review panel convened by the NRC concluded that the NNI,

"does not have the essential elements of a research strategy" for addressing nanotechnology-related EHS research.[28] The NNAP strongly recommends that the NNI build on this NRC review as it further develops an effective cross-agency EHS research strategy.

6. ***Targeted funding***—In the past, discussions over how well the Federal Government's nanotechnology effort is doing in the EHS realm have focused, to a great extent, on how much money the government is spending in this area. This focus has led to apparent discrepancies between NNI-identified investment in EHS-related research and independent assessments. In 2008, the GAO concluded that only 80 percent of research projects identified by the NN I as being relevant to EHS actually were relevant in this way.[29] Similarly, the 2009 National Research Council review of the NNI EHS research strategy concluded that in regard to nanotechnology EHS research, "the relevance of FY 2006 research projects to the research needs is generally overstated."

The NNAP acknowledges the importance of adequate funding and appropriate accounting of nano- technology EHS research. That said, the NNAP suggests that appropriate and targeted funding for strategic nanotechnology EHS research is more important than absolute dollar amounts. To ensure that emerging EHS issues are addressed effectively and in a way that yields useful information for regulators and policymakers, the NNI needs to help the scientific community establish a substantial core of exploratory research into biological and environmental interactions with nanomaterials. In addition, the Federal Government needs to ensure sufficient funds are available to mission-driven agencies to address specific issues that are arising.

A number of reports—including the NNI's 2008 EHS research strategy[30]—have identified specific knowledge gaps that need to be filled in order to underpin effective risk assessment and management. Although the NNI has no budget or allocation authority, it is critical that mechanisms are identified and developed to ensure agencies such as NIOSH, EPA, NIEHS, and FDA have access to sufficient funding to ensure these knowledge gaps are filled. Recent budget requests for nanotechnology EHS research indicate that moves are being made to ensure a more appropriate targeting of research funding. Within a record request for nanotechnology EHS funding of $116.9 million for FY 2011, three times the equivalent budget request of five years ago, agencies such as NIOSH, EPA, FDA, and CPSC are in line to receive *substantial funding increases for nanotechnology EHS research. Significantly, this will be the first time that FDA and CPSC will have had a specific allocation of funds to cover nanotechnology, a welcome move and one that the NNAP hopes is sustained over a number of years.

Recommendations

Recommendation 4-1. Risk Identification

The NSET Subcommittee's NEHI working group should develop clear principles to support the identification of plausible risks associated with the products of nanotechnology.

These principles should be high-level and should underpin effective problem formulation in crafting approaches to addressing emergent nanotechnology-related risks, which are defined here as risks that arise in unanticipated and poorly understood ways by virtue of the size and shape of the materials. Such risks thus transcend established approaches to risk assessment and management.

These principles should make it possible to differentiate those products of nanotechnology that are unlikely to present significant emergent risks from those that are more likely to do so. These principles should clarify that plausible risks depend both on hazard *and* a real likelihood of exposure, and support strategic action toward the goal of increasing certainty and predictability when addressing the safety of nanotechnology-related products. The principles should be made available to, and be suitable for use by, nanotechnology developers, users, and regulators.

Recommendation 4-2. Strategic Planning

The NSET Subcommittee's NEHI working group should further develop and implement a cross- agency strategic plan that links EHS research activities with knowledge gaps and decision-making needs within government and industry.

The strategic plan should respond to knowledge gaps identified in previous assessments of research needs, including the 2008 NNI EHS research strategy and the 2009 NRC review of the strategy.[31] In particular, the plan should ensure an appropriate balance between research addressing the release of engineered nanomaterials, human exposure to such releases, and the biological impacts of these materials following exposure. In addition, the strategic plan should support exploratory research that leads to new knowledge while ensuring a robust program of targeted research that addresses identified issues in a timely manner.

Mechanisms for closer integration and partnering between agencies that fund exploratory and applied research should be explored. The N NAP also encourages the formation and support by the NN I of investigator networks to facilitate the development of harmonized research methodologies. The strategic plan should be coordinated with the broader nanotechnology EHS research strategy currently being developed by the NRC. In the near term, the development and implementation of NNI's revised plan should be expedited and should not await the release of the NRC's strategy. The plan should also be reviewed on a regular basis—both internally and by nongovernment experts—and be updated when necessary.

RECOMMENDATION 4-3. ORGANIZATIONAL CHANGES

The NSET Subcommittee and OSTP should foster administrative changes and communications mechanisms that will enable the NNI to better embrace the EHS issues associated with nanotechnology research, development, and commercialization.

- The NSET Subcommittee co-chairs should assign an individual to NNCO to oversee I agency efforts that address nanotechnology EHS.
- OSTP and the NSET Subcommittee should expand the charter of the NEHI working group to enable the group to address cross-agency nanotechnology-related policy issues more broadly.
- The NSET Subcommittee should explore mechanisms that enable the NEHI working group to more effectively receive input and advice from nongovernment experts in the field of emergent risks.

The EHS-related point person within NNCO should constitute a liaison for stakeholders within and outside government on nanotechnology EHS issues, be able to interact with agencies at a senior level, and have a keen understanding of and experience with risk research and its relevance to decision making. This EHS leader within the NNI should be accountable for cross-agency actions toward the responsible development of nanotechnology-based products and processes, including development and implementation of the nanotechnology EHS strategic plan.

In this respect, the NSET Subcommittee's NEHI working group should not be bound by the NNI definition of nanotechnology. Instead, the NSET Subcommittee's NEH I working group should be encouraged to develop a risk-focused scope that encompasses nanoscale materials and products that potentially lead to new risks and/or challenge current approaches to risk assessment, management, and oversight.

In particular, the value of establishing a Federal Advisory Committee that would enable experts and stakeholders in the field of nanotechnology EHS to provide the NN I with expert advice, input, and information on a regular basis should be evaluated. The NNAP feels that such a committee could strengthen considerably the robustness and relevance of Federal actions in addressing nanotechnology EHS, while reducing the likelihood that cross-agency activities will fail to match up with stakeholder expectations.

RECOMMENDATION 4-4. INFORMATION RESOURCES

The NSET Subcommittee's NEHI working group should develop information resources on crosscutting nanotechnology EHS issues that are relevant to businesses, health and safety professionals, researchers, and consumers.

These information resources could include, but should not be limited to, the use and safety of nanomaterials in consumer products; "safety by design" approaches to developing these materials and products; risk assessment and risk management of nanotechnology-based products; and life-cycle approaches to responsible nanomaterial use.

Ten years ago, the United States blazed a new trail for national technology policy in its formation of, and investment in, the NNI. Its vision of new technologies and industries made possible by controlling matter on the nanoscale is fast becoming a reality. However, innovation alone is not enough for catalyzing a new economy in the 21st century. It also requires an early and honest evaluation of the possibilities that these same innovations could inadvertently cause harm. The public's and policymakers' perceptions of such risks will shape the commercialization path of emerging nanotechnology-based industries perhaps even more than the brilliance of the innovations. The NNI should leverage its international leadership and proven track record in research coordination in this area. Its ability to promote EHS research, coordinate research activities nationally and internationally, and speak to all nanotechnology stakeholders is singular among the agencies and is exactly what is required for this young industry.

V. Nanotechnology beyond 2010

Chapter Summary

The NNI was established 10 years ago with an optimistic and sweeping vision in which virtually every category of technology would, in time, be transformed as scientists and engineers became ever more adept at engaging and controlling matter on the nanoscale. In the years since, researchers have taken many steps toward realizing that vision in a range of areas that include materials, information technology, health care, energy, and national security. Over the next 10 years, various nanotechnologies that have been under development are likely to find their places in society and new ones will emerge from the core of fundamental research that underlies nanoscience and nanoengineering. The NNI's role in the coming years will need to coevolve with the maturing nanotechnology era.

In its January 2000 press release announcing the NNI, the White House included a visionary list of potential long-term outcomes of the initiative:

- Shrinking the entire contents of the Library of Congress in a device the size of a sugar cube through the expansion of mass storage electronics to multi-terabit memory capacity that will increase the memory storage per unit surface a thousand fold.
- Making materials and products from the bottom up, that is, by building them up from atoms and molecules. Bottom-up manufacturing should require less material and pollute less.
- Developing materials that are 10 times stronger than steel, but a fraction of the weight for making all kinds of land, sea, air, and space vehicles lighter and more fuel efficient.
- Improving the computer speed and efficiency of minuscule transistors and memory chips by factors of millions.
- Using gene and drug delivery to detect cancerous cells by nanoengineered magnetic resonance imaging (MRI) contrast agents or target organs in the human body.

- Removing the finest contaminants from water and air to promote a cleaner environment and potable water.
- Doubling the energy efficiency of solar cells.

This is a list that reflects perceived societal needs and objectives at the turn of the millennium. Looking back at the last decade of the NNI and its accomplishments—both anticipated and unexpected—the NNAP believes it can gauge its success and suggest a vision of the future.

Extending the Capabilities of Information Technology

The January 2000 press release mentioned an audacious goal of "shrinking the entire contents of the Library of Congress in a device the size of a sugar cube." The writer apparently envisioned some leapfrog technology that goes far beyond the hard disk drive. In the past decade, that old technology, based on reading and writing digital information on a spinning magnetic disk, has continued to improve, but today there is more excitement about reading and writing information in devices that have no mechanical moving parts. We now take for granted that countless books, songs, pictures, and movies can be stored on silicon chips small enough to fit inside cameras, cell phones, music players, and many other consumer products.

Each memory cell on these chips is a silicon field effect transistor (FET) of a particular design that has been around for many decades. This means that today's rapid progress in information storage is the result of industrial investment in R&D that has been driving incremental improvements in the miniaturization of an old solid-state memory technology. Experts agree that this technology is approaching its physical limits, but new and different solid-state memory technologies are already being tested in the market. Phase-change memory cells and magnetic random access memory cells are built from materials other than silicon and involve operating principles different from those of the transistor. Thus, further progress in solid state memory is assured, but only because the seeds of that progress had been sown by way of fundamental research done many decades ago and because the means of commercialization as well as end markets for these products have long been in place. The NNI has not yet had time to make a significant impact on these developments in memory technology, but it is poised to carry the baton forward toward the ultimate limits of information storage.

Perhaps even more important for the future of information technology, the NNI is helping to focus researchers on a more difficult problem: the search for a new device for *processing* information, as opposed to storing it. In contrast to the emerging memory devices, there is no new device waiting in the wings to replace the FET for digital logic. Yet economic constraints on power dissipation and handling of heat have halted improvements in computing speed for the last five years. The individual devices on chips continue to get smaller and cheaper, but the chips generate more heat. The transistors are unable to run faster because it is technically too difficult to cool them in affordable ways. That is why manufacturers now tout the number of cores (microprocessors) they can integrate on a single silicon chip, rather than the ever-increasing processor clock speeds that they used to rely on for marketing. Unless the FET is re-invented or replaced by some revolutionary "new switch" for digital logic in the next few years, the information technology industry will lose much of its dynamism.

To stave off such a loss, leading microelectronics manufacturers in the United States have been pooling their financial resources, through a consortium known as the Semiconductor Research Corporation, to promote forward-looking university research that could provide the foundation for future generations of electronics technologies. As that tradition of collaboration moves further into the nanotechnology era, these companies embarked several years ago on a new research thrust, called the Nanoelectronics Research Initiative (NRI)[32]. In building this initiative, they built on research infrastructures and collaborative frameworks forged under follow the NNI.

Today, the NRI is a superb example of industry-university cooperative research, involving more than 30 top universities in the United States. Many of the research projects are organized around four multi-university centers that are funded partly by the State governments of California, New York, Texas, and Indiana. Other funding goes to various NSF NSECs. NSF and NIST each solicit additional research proposals and support the top proposals through the NNI. Individual research projects focus on new device concepts, particularly those that promise to break the power dissipation bottleneck that is limiting computing speed.

It all looks straightforward in hindsight: companies pooling resources to encourage pre-competitive university research in the hope of revitalizing their industry, state governments promoting regional development of R&D talent and infrastructure, and Federal funding agencies investing in forward- looking research that is in the national interest. But NNI played a catalyzing role. It would have been far more difficult to get this going without the NNI, which highlighted the long-term potential of nano- technology investment for State governments and helped to steer some NSF funding toward NSECs and other relevant university-based research.

By pooling their resources, some forward-looking companies have been able to leverage some of this research capability in directions critical to the future of information technology. The impact of their investment has been magnified, and the odds of a big return on that investment in the coming decades have been substantially increased. Exploration and development of new devices for information technology will remain a grand challenge in the coming decade and the NNI is in a position to play important orchestrating roles.

Health Care in the 21st Century

That visionary White House press release a decade ago pointed to the use of designed nanostructures for diagnosis and treatments of cancer. "Within three years, researchers and program managers at the National Cancer Institute were contemplating "game changing" ways of taking on cancer. Based on promising research results going back to the late 1 990s, these NCI personnel began to create a nanomedicine program with a focus on cancer. With an eye on going beyond basic research toward nanomedicine technologies, the NCI recruited scientists with physical and chemical backgrounds along with clinicians to discuss how nanotechnology might help physicians detect and quash cancer.

Those involved in this organizational development credit the NNI with playing an important role in convincing NCI to launch its first large-scale program in nanotechnology, the NCI Alliance for Nanotechnology in Cancer.[33] The mere existence of the NNI demonstrated to the decision makers at NCI that the highest level of government was supporting a national, multi-agency effort in nanotechnology. The NNI's presence made it clear that there was Presidential support for a program in nanotechnology in the health sector and it showed that the NCI would not be out on its own with a nanotechnology program.

Proving that point, NNI leadership assisted the NCI Nanotechnology Alliance in creating multi-agency programs and facilities, including the Nanotechnology Characterization Laboratory (NCL)[34] with NIST, FDA, and other partners. On its Web site, the NCL describes its role as "a national resource and knowledge base for all cancer researchers to facilitate the regulatory review of nanotechnologies intended for cancer therapies and diagnostics."

At the moment, advances in nanomedicine are unfolding along several lines, among them analysis of medical samples, imaging, and therapeutics. The ability to measure multiple types of proteins, DNA, RNA, and other telling molecules, in ever tinier samples of blood and other tissues and in ever shorter times, is leading to medically-relevant data sets of unprecedented richness and value. The combination of technologies, such as microfluidics for moving liquids between tiny chambers on specialized diagnostic chips and nanotechnology-enabled methodologies for sensing molecules, are leading to tools for cheaply and rapidly making thousands of molecular measurements on single drops of blood and other biological fluids such as saliva.

By combining these kinds of tools with low-cost genomic sequencing, physicians will have new means for detecting, monitoring, and making decisions on how to treat cancer and other diseases. Instead of measuring one or two proteins, it will become possible to multiplex thousands of measurements to characterize even the subtly different disease pathways that occur in individual patients. The ability to perform diagnostic measurements and also monitor the outcomes of treatments in near real time will guide physicians in their quest to personalize the selection of therapies and to choose the best time course of administration for the most effective treatment. Because FDA requirements for the approval of diagnostics are not as elaborate as for *in vivo* imaging agents or therapeutics, diagnosis and treatment monitoring are areas of nanomedicine that will most likely be the first to provide "game changing" technologies for managing human diseases.

Another exciting trajectory in nanomedicine is the development of more sensitive and selective imaging methodologies. Imaging agents based on newly designed nanoparticles can reveal the presence of cancer cells far more readily than can molecular imaging agents that have been state of the art. The ability to fabricate nanostructures that act cooperatively to enhance the sensitivity of MRI and other imaging methods, combined with targeted delivery of these assemblies to increase concentrations at sites of disease, are making medical imaging ever more capable. These new imaging techniques go beyond anatomy to the molecules involved in disease. For example, researchers working on targeted nanoparticle-based imaging techniques expect that in the future, these new methods will enable them to scan the entire body at once for the presence of cancer that has spread from a primary tumor. Even more importantly, these nanoparticle technologies will enable physicians to classify the molecular signatures of each tumor, which is the sort of information they need to tailor therapeutic regimens.

Yet another promising direction in nanomedicine for the next decade is the development of new therapeutics that target diseased cells and tissue in more precise fashions. Medicines with such selectivity should bring with them far fewer side effects. For example, RNA interference refers to the use of small RNA molecules that can selectively inhibit the production of proteins in a very selective manner. The scientists who unraveled the molecular basis of this phenomenon shared the 2006 Nobel Prize for Medicine or Physiology. Nanotechnology now is playing a significant role in enabling RNA interference-based therapeutics for cancer treatments. By attaching these RNA molecules to nanoparticles designed to seek out and latch onto melanoma tumor cells, researchers have shown they can inhibit the production of a particular protein in a patient's cancer cells. Since RNA interference can be exploited to attack virtually any disease-associated protein, this phenomenon represents a therapeutic strategy that should have broad clinical implications. It should be pointed out that because of the significant amount of pre- approval studies required by the FDA, developing nanomedicines such as RNA-nanoparticle complexes is the most costly and most long-term endeavor in the nanomedicine context.

Beyond Steel: High Strength Materials

The White House press release in 2000 envisioned development of "materials that are 10 times stronger than steel, but a fraction of the weight for making all kinds of land, sea, air and space vehicles lighter and more fuel efficient." By that year, measurements of the stiffness and strength of CNTs, which are carbon atoms arranged in a chicken-wire-like lattice (graphene) and rolled up into minuscule seamless tubes, inspired politicians and engineers alike to imagine new high-strength, but low-weight materials. Single- walled carbon nanotubes (SWNTs) are nearly 10 times as strong as steel. A more important property of a material when it comes to applications needing lightweight structural materials is the ratio of strength to weight. For CNTs, when measured along their long axis, that ratio is about 20 times that of steel.

Reducing the weight of structural materials offers significant payoffs. A study released in 2002 by the National Academies, entitled *Effectiveness and Impact of Corporate Average Fuel Economy (CAFÉ) Standards,*[35] found that reducing the weight of a vehicle by 20 percent percent results in a 15 percent percent reduction in fuel consumption. Given the much greater strength-to-weight ratio of SWNTs compared to steel, nanostructured composite materials could allow designers to reduce the weight of structural components without compromising stiffness and other properties needed to maintain passenger safety.

It did not take long before an obstacle to realizing the technological promise of CNTs became evident. Conventional methods of fabricating fiber-reinforced composite materials depend on the ability to disperse fibers into uniform composites that are simultaneously stiff, strong, and *tough.* (Toughness is a measure of a material's resistance to crack propagation). For CNTs, which in this context constitute novel and unprecedentedly small fibers, the critical challenges included uniformly dispersing the nanotubes within a polymer or other matrix to enable sufficiently dense loadings and tailoring interfaces to control adhesion with the matrix to assure effective distribution of stresses. Engineering the surfaces of carbon nanotubes is particularly challenging because their surfaces are relatively inert and surface treatments to improve adhesion can damage them.

In the last few years, a way has been found to completely sidestep the loading problem, at least in some applications. With a novel process, materials scientists and engineer have managed to make and spin nanotubes directly into fibers and even cloth. CNT-based composites made with such starting materials are now being developed and tested for a wide variety of applications (See the case study on Nanocomp, page 34). Early applications address limited markets such as strong lightweight electrical wiring for satellites and aircraft, where weight reduction is crucial. However, the manufacturing processes appear to be scalable to large volumes at much lower cost, potentially making the material practical for large vehicles.

This is merely a glimpse of what could become possible with advanced composite materials. Nature's own nanocomposites, among them bone, shell, and wood, are simultaneously strong, light, and tough, even though they are composed from the somewhat limited set of materials available through biological evolution. The secret to these enviable combined properties is that natural materials are organized hierarchically at the nano, micro, and larger scales to achieve orders-of-magnitude increases in, for example, strength, and toughness compared to their constituent phases. Wood is a cellular composite with four levels of structure spanning the molecular to the macro scale. Its stiffness and strength per unit weight are comparable to steel, and its toughness is 10 times that of a conventional fiber-reinforced composite with a comparable loading of fibers. A challenging vision for the next 10 and 20 years is to emulate proven natural designs in manufacturable architectures where the ultimate properties of engineered nanoconstituents can be fully realized and enhanced.

Energy and the Environment

The world demand for energy is expected to double by the year 2050. The primary challenge is not how to meet this demand, but how to do so in an environmentally acceptable manner. This is where nanomaterials will come in; they will play a large role in creating sustainable, environmentally acceptable energy and fuels. Nanostructured materials already underlie significant advances in energy storage, transmission, and transformation in the form of, for example, lithium ion batteries, heat pumps, and photovoltaic cells, respectively.

Probably the most crosscutting area of science and technology for the energy and environmental sectors is catalysis, a field based on structures and phenomena at the nanoscale. Although catalysis is already a commercial success story in numerous energy and environmental applications—notably in the efficient catalytic "cracking" of petroleum into the smaller molecules of liquid fuels and in the catalytic converters on automobiles—there remains a critical need to develop new catalytic materials. Catalysts have been described in a NSF report as "the engines that power the world at the nanometer length scale."[36] Additionally, that report goes on to provide a grand challenge for catalysis scientists and engineers: "To control the composition and structure of catalytic materials over length scales from one nanometer to one micron to provide catalytic materials that accurately and efficiently control reaction pathways."

Future clean energy technologies such as the conversion of sunlight to fuels, fuels from biomass, and the splitting of water into hydrogen and oxygen will only be achievable via breakthroughs in catalysis. At this time, the understanding of how light interacts with catalytic materials lags significantly behind the knowledge on the interactions of molecules with catalytic materials. Advances in the fundamental physics and chemistry underlying catalysis, along with molecular self-assembly methods where cleverly designed precursors automatically assemble into catalytic nanostructures, should enable the design and fabrication of photocatalysts with improved efficiency in light absorption and/or conversion.

There is a history of successfully developing catalytic materials into commercial processes that have had large impacts on the energy and fuels sectors. New approaches with nanomaterials and systems that include them are likely to follow suit. Investments in nano-based catalytic science and technology are likely to have major payoffs because the pathways and associated risks for translation of fundamental discoveries in catalysis into commercial technology are established. According to a DOE Basic Energy Sciences Workshop, "Basic Research Needs: Catalysis for Energy,"[37] held in 2007, a 10-year grand challenge for realizing the full potential of catalysis for energy applications is the development of a profound understanding of catalytic transformations. That knowledge will become the foundation for designing and fabricating future catalysts with atom-by-atom precision and for converting reactants to products with molecular precision. Addressing this grand challenge should be a major priority of the NNI in the coming decade.

National Security

The events of 9/11, deaths from the subsequent anthrax attack, threats to encryption and cybersecurity, and the evolving complex socio-political climate have created a new context for national security during the last decade. During the next decade, nanotechnology has the potential to improve national security in many respects.

In strictly military contexts, developments in nanotechnology are expected to provide soldiers with smart, responsive, lightweight field technologies like adaptive camouflage, self-healing armor, wound- healing medicines, self-decontaminating clothing, and sensor and communication systems to allow complete situational awareness. Nanostructured quantum computing architectures based on control of the state of electron spin are envisioned to enable new levels of encryption and cybersecurity.

Nanotechnology development also is intimately linked with other security issues, among them energy production, our dependence on foreign oil and other resources, the global environment, and world health. All of these have security implications. Consider water. Scarcity of this fundamental resource could lead to international conflicts, and the availability of clean water is crucial to world health. Nanotechnologies are expected to enable the development of more efficient and less expensive ways to purify water. Specific technologies in this context could include nanostructured adsorbents, decontaminants, and filters.

Another way that nanotechnology is coupled to national security is in the detection of radiological, explosive, chemical, and biological agents. Sensitive, fast, stable sensors capable of operation in complex environments are needed for detection of a spectrum of potential threats. It is generally argued that nanoscale sensors, such as nanowire-based field effect transistors, are inherently more sensitive than microscale systems. Taking a cue from the immune system, which can rapidly detect tiny quantities of a specific molecule, researchers have begun integrating engineered biological nanocomponents such as antibodies, nucleic acids, and proteins with nanoscale sensors to develop devices that can detect specific threats. Some of the challenges of threat detection are shared with those of medical diagnostic systems such as detection of tiny amounts of analytes, multiplexing to avoid false positives, and simplicity of operation. During the next decade, it is likely that some nanotechnology strategies under current development for biological assays will also contribute to national defense.

A Vision for the Next 10 Years

In its first 10 years, the NNI has promoted progress on scientific problems of great importance to society. It has increased dramatically the investment in instrumentation, research infrastructure, and expertise. Looking forward 10 years, the NNAP sees a program that builds on these strengths. Progress must continue along lines of research such as those sketched above, but new avenues of investigation will spring up based on changing societal needs and unpredicted discoveries of a transformative and revolutionary nature. A vibrant and effective NNI will have the following attributes:

- ***Basic research will remain a critical component of the research portfolio.*** N NI will continue to provide an organizational structure that promotes crosscutting research that stands to enhance our economic competitiveness. New fundamental discoveries will continue to refresh our ideas of what is possible and provide the foundation for new initiatives.
- While basic research continues, there will be increasing focus on integration of components and processes that lead to commercialization. For example, integration of nanotechnology-enabled diagnosis, imaging, and therapy will provide superior new methodologies for the management of cancer. Low-cost, sensitive, rapid diagnostics combined with whole-body imaging will allow earlier detection of cancer. Potent and selective therapies will be matched to each patient's disease characteristics and in vivo diagnostics and imaging will guide the course of treatment. Research within the auspices of N NI, coordinated with other appropriate research initiatives, will strive to tightly integrate advances in nanotechnology into an individualized cancer management strategy.
- The NNI will play a key role in several Signature Initiatives leveraging targeted interagency efforts to address grand challenges. The FY 2011 budget accommodates three Signature Initiative platforms for nanoscience and nanotechnology: Nanotechnology Applications for Solar Energy, Sustainable Nanomanufacturing, and Nanoelectronics for 2020 and Beyond. These and other bold initiatives will be chosen for their potential to improve quality of life, protect the environment, create jobs, and engage a new generation of scientists and engineers. Among some of the other Signature Initiatives on which the NNI could partner with industry, academia, and the relevant agencies are regenerative medicine, catalysis, food safety, and threat detection. These initiatives, complemented by a vibrant educational program which focuses on the potential of nanotechnology to help solve societal problems, will excite the imaginations of young people and draw the next generation of scientists and engineers into the field. For example, a new Signature Initiative

to push catalysis forward may make possible the efficient splitting of water into hydrogen (fuel) and oxygen, or the use of carbon dioxide as a chemical feedstock. Translating such advances into industrial scale processes will be accelerated if the NNI continues to fund research on scientific issues critical to scale up.

- ***The balance of NNI programs will continue to evolve.*** In the coming years, for example, there will be a stronger focus on fundamental issues related to EHS. What are the structural and chemical features of nanoparticles that determine transport through, concentration in, and interactions with the human body? The effort to address those questions will be coordinated across NIH, NIST, EPA, and other relevant agencies. The increased focus on integration and system demonstrations will demand an increased focus on research that is relevant to assembly and eventual manufacture of such systems. Exploration of relatively simple self-assembly processes that are suitable for batch fabrication of relatively simple systems will be supplemented by research into fabrication of complex structures and systems with many interacting nanoscale components. Research will therefore focus on increasingly sophisticated processes and systems for directed and templated self-assembly. There will also be a component of manufacturing research focused on new processes for precision patterning, looking beyond the optical and e-beam lithographic processes driven by the microelectronics industry. Novel means of fabrication with ever finer precision should be the focus of new investments in nanomanufacturing, including modeling and simulation, metrology tools, and the merging of self-assembly with lithography to achieve large-scale predictable placement of nanoscale components.

Appendix A. Statement of Task

President's Council of Advisors on Science and Technology Review of the National Nanotechnology Initiative

The 21st Century Nanotechnology Research and Development Act of 2003 (Public Law 108-153) calls for a National Nanotechnology Advisory Panel to periodically review the Federal nanotechnology research and development program known as the National Nanotechnology Initiative. The President's Council of Advisors on Science and Technology is designated by Executive Order to serve as the NNAP.

The study intends to answer the following questions from The 21st Century Nanotechnology Research and Development Act of 2003 (Public Law 108-153):

1. What are the trends and developments in nanotechnology science and engineering, as they relate to the NNI and generally?
2. Please describe NNI's progress in the last two years in implementing the NNI Program.
3. Does the NNI Program need to be revised? If so, how?
4. Is the balance correct among the components of the NNI Program, including funding levels for the program component areas?

5. Have the component areas, priorities, and technical goals helped to maintain U.S. leadership in nanotechnology?
6. Has the management, coordination, implementation, and activities of the program being been carried out appropriately over the last two years? How could improvements be made?
7. Have societal, ethical, legal, environmental, and workforce concerns been adequately addressed by the Program during the last two years?

The study will address additional questions:

8. What should be the goals, priorities and platforms for the NNI for the next five years in science, biomedical research, nano-renewables, and nano-electronics?

Environment, Health, and Safety (EHS):

9. Has NNI established an acceptable strategy to address appropriate EHS priorities?
    - Is it effective?
    - Is it filling knowledge gaps?
    - Does it have an implementation and review plan?
    - Does it support evidence-informed decision making?
10. Is NNI appropriately invested in EHS research to address the priorities?

11. Has NNI implemented an approach toward achieving the goals of the strategic plan that leverages the strengths of Federal agencies?
    - Have experts and other stakeholders outside the Federal Government been engaged effectively in addressing nano-EHS issues?
    - Have experts had input to the research strategy?
    - Have industry, consumers and other stakeholders been engaged in identifying and addressing EHS issues?
    - Have State agencies been engaged in addressing EHS issues?
12. Has collaboration between Federal agencies been effective at addressing nano-EHS issues?
    - Have cross-agency actions been coordinated or conflicting?
    - Have efforts to ensure interagency collaboration added value to the NNI?
    - Is there clear evidence of cross-agency collaboration having a significant impact on agency decisions and strategies?
13. Have Federal agencies collaborated effectively with international partners on addressing nanoEHS issues?
    - Is there a clear and substantial awareness of international developments and actions?
    - Are Federal agencies engaged with international partners, and are these engagements coordinated across the Federal Government?
    - Is there clear evidence for the United States influencing international developments, and benefiting from international collaborations?
14. Are there mechanisms in place to ensure research into the EHS implications of nanotechnology leads to evidence-informed decision making?

- Is there a clear link between research and policy?
- Is research targeted to providing information regulators need to make informed decisions?
- Are there mechanisms to coordinate regulatory actions on nanotechnology across Federal agencies?

Nanotechnology Outputs

15. What are the trends and developments in nanotechnology science and engineering and product introduction since 2008 (the last PCAST review): in the United States, internationally, and at the level of individual States?
16. Is the United States the leader, in what areas, and by what metrics?
17. Has progress been made since the last review in the transfer of science to products and processes, in technology transfer, and in the development of enabling standards?
18. Does an adequate science base to facilitate transfer of nanotechnology exist, if so in what areas, if not where are the gaps?
19. Has progress been made since the last review in the transfer of science to products and processes, in technology transfer, and in the development of enabling standards?

Program Management

20. Has appropriate progress been made since the last PCAST review in managing and implementing the program?
    - Has an appropriate strategic plan been developed for NNI broadly, and does it include societal, ethical, legal and workforce objectives?
    - Do clear objectives exist to achieve the goals of the strategic plan?
    - Is the NNI appropriately balanced with regard to the components of the Program, including funding levels for the program component areas?
    - What is the status of program management as it relates to coordination and implementation of efforts across NNI?
21. Is there a need to modify the strategic plan that was in place at the last review to reflect changing priorities or developments in the NNI?
22. Has the investment in equipment infrastructure been adequate, and has there been some larger coordination among agencies in developing that infrastructure?
23. Have the investments in education and communication to the public been effective?
    - What has been learned?
    - What more should be done?
24. If the societal, ethical, legal and workforce objectives have not been well integrated into the program, how should NNI proceed in the future?
25. What is the status of program management as it relates to coordination and implementation of efforts across NNI?
    - Are there areas that require greater emphasis and coordination?

26. Does the NNI have the ability to quickly respond and adapt to new scientific/technological advances?
27. What is the impact of NNI beyond the creation of publications and patents?

APPENDIX B. STUDY DESIGN—REVIEW OF THE NATIONAL NANOTECHNOLOGY INITIATIVE

A Statement of Task was developed (see Appendix A)

- Requirements from Public Law 108-153 were included.
- Interviews (see Appendix B) were conducted.

Working Group members were selected and vetted.
Working Group members sorted into teams to complete the study:

- Program management
- Environment, health, and safety
- Outputs of NNI
- Vision for the future of nanotechnology

Weekly Working Group conference calls were held for administrative purposes to refine and conduct the review

Two in-person meetings were held for preparatory purposes to gather expert opinions (see Appendix C).

- January 19–20, 201 0, Washington, DC.
- February 18–19, 2010, Palo Alto, CA.

All invited experts were asked to complete a survey comprised of questions from the Statement of Task.

Additional data gathering and analysis was conducted at the request of the Working Group by analysts at the Science and Technology Policy Institute.

Recommendations were developed and a study report was written by Working Group members.

The NNI review report and its recommendations were presented to PCAST on March 12, 2010.

APPENDIX C. FEDERAL AGENCIES PARTICIPATING IN THE NNI

Federal Agencies with Budgets Dedicated to Nanotechnology Research and Development

Consumer Product Safety Commission (CPSC)

Department of Defense (DOD)
Department of Energy (DOE)
Department of Homeland Security (DHS)
Department of Justice (DOJ)
Department of Transportation (DOT, including the Federal Highway Administration, FHWA)
Environmental Protection Agency (EPA)
Food and Drug Administration (FDA, Department of Health and Human Services)
Forest Service (FS, Department of Agriculture)
National Aeronautics and Space Administration (NASA)
National Institute for Occupational Safety and Health (NIOSH, Department of Health and Human Services/Centers for Disease Control and Prevention)
National Institute of Food and Agriculture (NIFA, Department of Agriculture)[38]
National Institutes of Health (NIH, Department of Health and Human Services)
National Institute of Standards and Technology (NIST, Department of Commerce)
National Science Foundation (NSF)

Other participating agencies

Bureau of Industry and Security (BIS, Department of Commerce)
Department of Education (DOEd)
Department of Labor (DOL)
Department of State (DOS)
Department of the Treasury (DOTreas)
Director of National Intelligence (DNI)
International Trade Commission (ITC)
Nuclear Regulatory Commission (NRC)
U.S. Geological Survey (USGS, Department of the Interior)
U.S. Patent and Trademark Office (USPTO, Department of Commerce)

APPENDIX D. LIST OF ACRONYMS

CBI	confidential business information
CNT	carbon nanotubes
CPSC	Consumer Products Safety Commission
CR-BSI	catheter-related blood stream infections
DOD	Department of Defense
DOE	Department of Energy
EHS	environment, health, and safety
EU27	27 nations of the European Union
FDA	Food and Drug Administration
FET	field effect transistor
GAO	Government Accountability Office
GDP	gross domestic product

GNP	gross national product
ICON	International Council on Nanotechnology
IND	Investigational New Drug Application
LAD	Luer-activated device
M&A	mergers and acquisitions
MRSEC	Materials Research Science and Engineering Centers
NCI	National Cancer Institute
NEHI	Nanotechnology, Environmental, and Health Implications
NIEHS	National Institute of Environmental Health Sciences
NIH	National Institutes of Health
NIOSH	National Institute for Occupational Health and Safety
NISE	Nanoscale Informal Science Education
NIST	National Institute of Standards and Technology
NNAP	National Nanotechnology Advisory Panel
NNCO	National Nanotechnology Coordination Office
N NI	National Nanotechnology Initiative
NRC	National Research Council
NRI	Nanoelectronics Research Initiative
NSEC	National Science and Engineering Centers
NSET	Nanoscale Science, Engineering, and Technology Subcommittee of the NSTC
NSF	National Science Foundation
NSTC	National Science and Technology Council
OECD	Organisation for Economic Co-operation and Development
OMB	Office of Management and Budget
OSTP	Office of Science and Technology Policy
PCA	Program Component Area
PCAST	President's Council of Advisors on Science and Technology
R&D	research and development
SBIR	Small Business Innovation Research
SOT	Statement of Task
STTR	Small Business Technology Transfer
SWCNT	single-walled carbon nanotube
TIP	Technology Innovation Program
VC	venture capital

End Notes

[1] National Science and Technology Council. February 2010. The National Nanotechnology Initiative –Supplement to the President's FY 2011 Budget, p. 5. Available at www.nano.gov/NNI_2011_budget_supplement.pdf.
[2] http://www.nano.gov/NNI_Strategic_Plan_2007.pdf
[3] www.nano.gov/html/res/FINAL_PCAST_NANO_REPORT.pdf
[4] www.nano.gov/PCAST_NNAP_NNI_Assessment_2008.pdf
[5] National Science and Technology Council. February 2010. The National Nanotechnology Initiative –Supplement to the President's FY 2011 Budget, p. 5. Available at www.nano.gov/NNI_2011_budget_supplement.pdf.

[6] Working groups on (1) Global Issues in Nanotechnology, (2) Nanotechnology Environmental and Health, (3) Nanomanufacturing, Industry Liaison, & Innovation, (4) Nanotechnology Public Engagement & Communications.
[7] Lux Research, "Nanotechnology State of the Market Q1 2009," 2009, Section 4.1.
[8] Lux Research, "Nanotechnology State of the Market Q1 2009," 2009.
[9] Lux Research.
[10] Lux Research, "The Recession's Ripple Effect on Nanotech," 2009.
[11] As listed in the Science Citation Index.
[12] These 12 are identified by Leydesdorif [*Scientometrics, Vol. 76, No. 1 (2008) 159–167*] as: *Advanced Drug Delivery Reviews, Advanced Materials, Chemistry of Materials, Current Nanoscience, Fullerenes Nanotubes and Carbon Nanostructures, Journal of Nanoparticle Research, Journal of Nanoscience and Nanotechnology, Journal of Physical Chemistry B, Microfluidics and Nanofluidics, Nano Letters, Nanotechnology*, and *Nature Materials*. They do not include *Science*, *Nature*, and Proceedings of the National Academy of Sciences, in which the most highly cited articles in nanoscience appear.
[13] Lux Research, "Nanotechnology State of the Market Q1 2009," 2009, Section 4.2.
[14] Lux Research, "Nanotechnology State of the Market Q1 2009," 2009, Section 4.2.
[15] Lux Research, "Nanotechnology State of the Market Q1 2009," 2009, Section 2.
[16] Lux Research, "The CEO's Nanotechnology Playbook," 2005.
[17] Lux Research, "Nanotechnology State of the Market: Stealth Success, Broad Impact." 2008.
[18] Op. cit.
[19] Lux Research, "2009 Nanotech Venture Capital: Healthcare and Life Sciences Provide Life Support," 2010.
[20] www.nano.gov/PCAST_NNAP_NNI_Assessment_2008.pdf
[21] www.nano.gov/NNI_2011_budget_supplement.pdf
[22] www.nano.gov/NNI_EHS_Research_Strategy.pdf
[23] www.epa.gov/oppt/nano/stewardship.html
[24] www.nano.gov/html/meetings/exposure/index.html
[25] www.characterizationmatters.org
[26] www.nanoehsalliance.org
[27] www.icon.rice.edu/virtualjournal.cfm
[28] www.nap.edu/catalog.php?record_id=12559
[29] www.gao.gov/products/GAO-08-709T
[30] www.nano.gov/NNI_EHS_Research_Strategy.pdf
[31] www.nap.edu/catalog.php?record_id=12559
[32] nri.src.org/member/about/default.asp
[33] nano.cancer.gov/
[34] ncl.cancer.gov/
[35] www.nap.edu/openbook.php?isbn=0309076013
[36] From NSF Workshop Report on "Future Directions in Catalysis: Structures that Function on the Nanoscale," NSF Headquarters, Arlington, VA, June 19-20, 2003, Davis, M., Chair, Tilley, Don, Co-Chair.
[37] www.er.doe.gov/bes/reports/files/CAT_rpt.pdf
[38] Section 7511 of the Food, Conservation, and Energy Act of 2008 (FCEA) established within the Department of Agriculture the National Institute of Food and Agriculture (NIFA) and transferred all authorities of the Cooperative State Research, Education, and Extension Service (CSREES) to NIFA not later than October 1, 2009.

In: Nanotechnology: Looking Ahead
Editor: Julian S. Moser

ISBN: 978-1-61761-785-0

Chapter 5

NANOTECHNOLOGY: THE FUTURE IS COMING SOONER THAN YOU THINK

Jim Saxton(R-NJ)

ABSTRACT

Enhanced abilities to understand and manipulate matter at the molecular and atomic levels promise a wave of significant new technologies over the next five decades. Dramatic breakthroughs will occur in diverse areas such as medicine, communications, computing, energy, and robotics. These changes will generate large amounts of wealth and force wrenching changes in existing markets and institutions.

This paper discusses the range of sciences currently covered by nanotechnology. It begins with a description of what nanotechnology is and how it relates to previous scientific advances. It then describes the most likely future development of different technologies in a variety of fields. The paper also reviews the government's current nanotechnology policy and makes some suggestions for improvement.

INTRODUCTION

In 1970 Alvin Toffler, noted technologist and futurist, argued that the acceleration of technological and social change was likely to challenge the capacity of both individuals and institutions to understand and to adapt to it.[1] Although the world has changed a great deal since then, few would argue that the pace of change has had the discontinuous effects that Toffler predicted. However, rapid advances in a number of fields, collectively known as nanotechnology, make it possible that Mr. Toffler's future has merely been delayed. In fact, some futurists now talk about an unspecified date sometime around the middle of this century when, because of the accelerating pace of technology, life will be radically different than at any prior time.

This paper discusses the range of sciences currently covered by nanotechnology. It begins with a description of what nanotechnology is and how it relates to previous scientific advances. It then describes the most likely future development of different technologies in a variety of fields. The paper also reviews the federal government's current nanotechnology policy and makes some suggestions for improvement.

WHAT IS NANOTECHNOLOGY?

A nanometer (nm) is one billionth of a meter. For comparison purposes, the width of an average hair is 100,000 nanometers. Human blood cells are 2,000 to 5,000 nm long, a strand of DNA has a diameter of 2.5 nm, and a line of ten hydrogen atoms is one nm.[2] The last three statistics are especially enlightening. First, even within a blood cell there is a great deal of room at the nanoscale. Nanotechnology therefore holds out the promise of manipulating individual cell structure and function. Second, the ability to understand and manipulate matter at the level of one nanometer is closely related to the ability to understand and manipulate both matter and life at their most basic levels: the atom and the organic molecules that make up DNA.

Nanotechnology can be viewed on a variety of levels. The U.S. National Nanotechnology Initiative defines nanotechnology as:

> "[T]he science, engineering, and technology related to the understanding and control of matter at the length scale of approximately 1 to 100 nanometers. However, nanotechnology is not merely working with matter at the nanoscale, but also research and development of materials, devices, and systems that have novel properties and functions due to their nanoscale dimensions or components"[3]

A joint report by the British Royal Society and the Royal Academy of Engineering similarly defined nanotechnology as "the design, characterization, production, and application of structures, devices and systems by controlling shape and size at nanometer scale."[4]

The application of nanotechnology can occur in one, two or three dimensions. Thus it includes the use of an oxygen plasma 25 atoms thick to bind a layer of indium phosphide to silicon in order to make a computer chip that uses lasers to transmit data at 100 times the speed of current communications equipment.[5] In two dimensions it includes the manufacture of carbon nanotubes only one nanometer in diameter that may be eventually reach several centimeters in length. In three dimensions it encompasses the manufacture of small particles no more than a few nanometers in any dimension that might be used as an ingredient in sunscreens or to deliver medicine to a specific type of cell in the body.

In a more general context nanotechnology can be seen as just the current stage of a long-term ability to understand and manipulate matter at ever smaller scales as time goes by. Over the last century, physicists and biologists have developed a much more detailed understanding of matter at finer and finer levels. At the same time, engineers have gradually acquired the ability to reliably manipulate material to increasingly finer degrees of precision. Although we have long known much of what happens at the nanolevel, the levels of knowledge implied by; 1) knowing about the existence of atoms, 2) actually seeing them, 3) manipulating them, and 4) truly understanding how they work, are dramatically different. The last two stages especially open up significant new technological abilities. At the nanolevel technology has just recently reached these stages.

Two examples indicate the significance of current research. Biologists have known about the basic building blocks of DNA since 1953, but until recently did not know the exact DNA sequence of a human being. This occurred in the last decade. Viruses were another mystery, but now scientists not only know the DNA sequence, they have used this knowledge to build a virus that assembles a battery.[6] As a second example, rather than just being able to see individual atoms with an electron microscope, scientists can now place a 20-nm indentation on a piece of material, creating a data storage system with the capacity to store 25 million printed textbook pages on a square inch chip.[7]

What makes work at the nanolevel more than just a natural progression of earlier work at the micro and macro levels of matter? For one thing the basic building blocks of matter and life occur at the nanolevel. Molecular chemistry, genetic reproduction, cellular processes, and the current frontier of electronics all occur on the nanolevel. Understanding how these processes work and, more importantly, being able to reliably manipulate events at this level in order to get specific outcomes, opens up the possibility of significant new advances in a wide variety of fields including electronics, medicine, and material sciences.

Second, the nanolevel represents the overlap between traditional physics and quantum mechanics. At this scale the physical, chemical, and biological properties of materials differ in fundamental ways from the properties of either individual atoms or bulk matter.[8] This makes the prediction of cause and effect relationships much more difficult and introduces phenomena such as quantum tunneling, superposition, and entanglement. As a result, material at the nanoscale can exhibit surprising characteristics that are not evident at large scales. For example:

- Collections of gold particles can appear orange, purple, red, or greenish, depending upon the specific size of the particles making up the sample.[9]
- Carbon atoms in the form of a nanotube exhibit tensile strengths 100 times that of steel and can be either metallic or semiconducting depending on their configuration.
- Titanium dioxide and zinc oxide, common ingredients in sun screen, both appear white when made of macro particles. But when the particles are ground to the nanoscale, they appear translucent.

The Progression of Nanotechnology

Why now? If it seems that nanotechnology has begun to blossom in the last ten years, this is largely due to the development of new instruments that allow researchers to observe and manipulate matter at the nanolevel. Technologies such as scanning tunneling microscopy, magnetic force microscopy, and electron microscopy allow scientists to observe events at the atomic level. At the same time, economic pressures in the electronics industry have forced the development of new lithographic techniques that continue the steady reduction in feature size and cost. Just as Galileo's knowledge was limited by the technology of his day, until recently a lack of good instrumentation prevented scientists from gaining more knowledge of the nanoscale. As better instrumentation for observing, manipulating and measuring events at this scale are developed, further advances in our understanding and ability will occur.

One leader in nanotechnology policy has identified four distinct generations in the development of nanotechnology products, to which we can add a possible fifth:[10]

Passive Nanostructures (2000-2005)

During the first period products will take advantage of the passive properties of nanomaterials, including nanotubes and nanolayers. For example, titanium dioxide is often used in sunscreens because it absorbs and reflects ultraviolet light. When broken down into nanoparticles it becomes transparent to visible light, eliminating the white cream appearance associated with traditional sunscreens. Carbon nanotubes are much stronger than steel but only a fraction of the weight. Tennis rackets containing them promise to deliver greater stiffness without additional weight. As a third example, yarn that is coated with a nanolayer of material can be woven into stain-resistant clothing. Each of these products takes advantage of the unique property of a material when it is manufactured at a nanoscale. However, in each case the nanomaterial itself remains static once it is encapsulated into the product.

Active Nanostructures (2005-2010)

Active nanostructures change their state during use, responding in predicable ways to the environment around them. Nanoparticles might seek out cancer cells and then release an attached drug. A nanoelectromechancial device embedded into construction material could sense when the material is under strain and release an epoxy that repairs any rupture. Or a layer of nanomaterial might respond to the presence of sunlight by emitting an electrical charge to power an appliance. Products in this phase require a greater understanding of how the structure of a nanomaterial determines its properties and a corresponding ability to design unique materials. They also raise more advanced manufacturing and deployment challenges.

Systems of Nanosystems (2010-2015)

In this stage assemblies of nanotools work together to achieve a final goal. A key challenge is to get the main components to work together within a network, possibly exchanging information in the process. Proteins or viruses might assemble small batteries. Nanostructures could self-assemble into a lattice on which bone or other tissues could grow. Smart dust strewn over an area could sense the presence of human beings and communicate their location. Small nanoelectromechancial devices could search out cancer cells and turn off their reproductive capacity. At this stage significant advancements in robotics, biotechnology, and new generation information technology will begin to appear in products.

Molecular Nanosystems (2015-2020)

This stage involves the intelligent design of molecular and atomic devices, leading to "unprecedented understanding and control over the basic building blocks of all natural and man-made things."[11] Although the line between this stage and the last blurs, what seems to distinguish products introduced here is that matter is crafted at the molecular and even atomic level to take advantage of the specific nanoscale properties of different elements. Research will occur on the interaction between light and matter, the machine-human interface, and atomic manipulation to design molecules. Among the examples that Dr. Roco foresees are "multifunctional molecules, catalysts for synthesis and controlling of engineered nanostructures, subcellular interventions, and biomimetics for complex system dynamics and control."[12] Since the path from initial discovery to product application takes 10-12 years,[13] the initial scientific foundations for these technologies are already starting to emerge from laboratories. At this stage a single product will integrate a wide variety of capacities including independent power generation, information processing and communication, and mechanical operation. Its manufacture implies the ability to rearrange the basic building blocks of matter and life to accomplish specific purposes. Nanoproducts regularly applied to a field might search out and transform hazardous materials and mix a specified amount of oxygen into the soil. Nanodevices could roam the body, fixing the DNA of damaged cells, monitoring vital conditions and displaying data in a readable form on skin cells in a form similar to a tattoo. Computers might operate by reading the brain waves of the operator.

The Singularity (2020 and beyond)

Every exponential curve eventually reaches a point where the growth rate becomes almost infinite. This point is often called the Singularity. If technology continues to advance at exponential rates, what happens after 2020? Technology is likely to continue, but at this stage some observers forecast a period at which scientific advances aggressively assume their own momentum and accelerate at unprecedented levels, enabling products that today seem like science fiction. Beyond the Singularity, human society is incomparably different from what it is today. Several assumptions seem to drive predictions of a Singularity[14]. The first is that continued material demands and competitive pressures will continue to drive technology forward. Second, at some point artificial intelligence advances to a point where computers enhance and accelerate scientific discovery and technological change. In other words, intelligent machines start to produce discoveries that are too complex for humans. Finally, there is an assumption that solutions to most of today's problems including material scarcity, human health, and environmental degradation can be solved by technology, if not by us, then by the computers we eventually develop.

Whether or not one believes in the Singularity, it is difficult to overestimate nanotechnology's likely implications for society. For one thing, advances in just the last five years have proceeded much faster than even the best experts had predicted. Looking forward, science is likely to continue outrunning expectations, at least in the medium- term. Although science may advance rapidly, technology and daily life are likely to change at a much slower pace for several reasons. First, it takes time for scientific discoveries to become embedded into new products, especially when the market for those products is uncertain. Second, both individuals and institutions can exhibit a great deal of resistance to change. Because new technology often requires significant organizational change and cost in order to have its full effect, this can delay the social impact of new discoveries. For example, computer technology did not have a noticeable effect on economic productivity until it became widely integrated into business offices and, ultimately, business processes. It took firms over a decade to go from replacing the typewriters in their office pools to rearranging their entire supply chains to take advantage of the Internet. Although some firms adopted new technologies rapidly, others, lagged far behind.

THE STRUCTURE OF NANOTECHNOLOGY

Nanotechnology is distinguished by its interdisciplinary nature. For one thing, investigations at the nanolevel are occurring in a variety of academic fields. More important, the most advanced research and product development increasingly requires knowledge of disciplines that, until now, operated largely independently. These areas include:

- *Physics* — The construction of specific molecules is governed by the physical forces between the individual atoms composing them. Nanotechnology will involve the continued design of novel molecules for specific purposes. However, the laws of physics will continue to govern which atoms will interact with each other and in what way. In addition, researchers need to understand how quantum physics affects the behavior of matter below a certain scale.
- *Chemistry* — The interaction of different molecules is governed by chemical forces. Nanotechnology will involve the controlled interaction of different molecules, often in solution. Understanding how different materials interact with each other is a crucial part of designing new nanomaterials to achieve a given purpose.
- *Biology* — A major focus of nanotechnology is the creation of small devices capable of processing information and performing tasks on the nanoscale. The process by which information encoded in DNA is used to build proteins, which then go on to perform complex tasks including the building of more complex structures, offers one possible template. A better understanding of how biological systems work at the lowest level may allow future scientists to use similar processes to accomplish new purposes. It is also a vital part of all research into medical applications.
- *Computer Science* — Moore's Law and its corollaries, the phenomena whereby the price performance, speed, and capacity of almost every component of the computer and communications industry has improved exponentially over the last several decades, has been accompanied by steady miniaturization. Continued decreases in transistor size face physical barriers including heat dissipation and electron tunneling that require new technologies to get around. In addition, a major issue for the use of any nanodevices will be the need to exchange information with them. Finally, scientific advances will require the ability to manage increasingly large amounts of information collected from a large network of sensors.[15]

- *Electrical Engineering* — To operate independently, nanodevices will need a steady supply of power. Moving power into and out of devices at that scale represents a unique challenge. Within the field of information technology, control of electric signals is also vital to transistor switches and memory storage. A great deal of research is also going into developing nanotechnologies that can generate and manage power more efficiently.
- *Mechanical Engineering* — Even at the nanolevel issues such as load bearing, wear, material fatigue, and lubrication still apply. Detailed knowledge of how to actually build devices that do what we want them to do with an acceptable level of confidence will be a critical component of future research.

Unfortunately, most of academia and the research community do not facilitate this type of multidisciplinary research. Work often tends to be compartmentalized into disciplines and subdisciplines with their own vocabularies. Research proposals are evaluated by experts within one area who neither understand nor appreciate developments in other fields. Young people coming into a field are usually rewarded for extending existing lines of research and take a risk if they try to look at the unexamined gaps between academic fields.

Yet in nanotechnology most of the great possibilities are precisely in these gaps. In 2002 the National Academy of Sciences listed several important areas for investment in nanotechnology. All of them involved interdisciplinary research.[16] The National Science Foundation is trying to encourage such research by awarding grants specifically for it.

With so many sciences having input into nanotechnology research, it is only natural that the results of this research are expected to have a significant impact on a similarly broad range of applications. Ray Kurzwiel labels these applications genetics, nanotechnology, and robotics (GNR),[17] to which one can add information technology (GRIN).[18] The National Nanotechnology Initiative has adopted the similar classification of nanotechnology, biotechnology, information technology, and cognitive science (NBIC).[19]

These sciences interrelate in a number of ways:

Nanotechnology — Nanotechnology often refers to research in a wide number of fields including the other three listed below. But in its limited sense it refers to the ability to observe and manipulate matter at the level of the basic molecules that govern genetics, cell biology, chemical composition, and the current and future generations of electronics. Researchers can then apply this ability to advance science in other fields. The broader definition of nanotechnology applies throughout most of this paper, but it is worth remembering that advances in other sciences depend on continued improvements in the ability to observe, understand, and control matter at the nanolevel. This in turn will require more accurate and less expensive instrumentation and better techniques for producing large numbers of nanodevices.

Biotechnology (Genetics) — Nanotechnology promises an increased understanding and manipulation of the basic building blocks underlying all living matter. The basic theory of genetic inheritance has been known for some time. But biologists do not fully understand the details of how life goes from a single fertilized egg with a full set of chromosomes to a living animal. Questions exist on exactly how the information encoded in DNA is transcribed, the role of proteins, the internal workings of the cell and many other areas. Basically DNA consists of a long string of four molecules; adenine, thymine, guanine, and cytosine. Since these molecules are read off in units of three (called codons), there are 64 possible combinations. Each combination corresponds to one of 20 amino acids. The amino acids in turn form proteins that fold in unique three dimensional ways and perform many of the functions within individuals cells. On a basic level, research is allowing us to tease out the genetic basis for specific diseases and in the future may reliably allow us to correct harmful mutations. But what would a full understanding of the genetic process give us? Could we develop DNA that uses a fifth and sixth molecule? Could the existing process be reprogrammed to code for more than 20 amino acids? To what extent is it possible to create brand new proteins that perform unique functions?

A better understanding of biological processes is obviously needed in order to deliver the health benefits that nanotechnology promises. But it is also important for many reasons outside of biology. Those used to traditional manufacturing techniques may at first have difficulty with the concept of building a product up from the molecular level. Biology offers a template for doing so. A single fertilized egg in the womb eventually becomes a human being; a system of incredible complexity from a simple set of instructions 2.5 nm in diameter. Scientists are hopeful that similar processes can be used to produce a range of other products.

Information Technology — Progress in information processing has depended on the continued application of Moore's law, which predicts a regular doubling of the number of transistors that can be placed on a computer chip. This produced exponential improvements in computing speed and price performance. Current computer technology is based on the Complementary Metal Oxide Semiconductor (CMOS). The present generation of computer chips already depends on features as small as 70 nanometers. Foreseeable advances in nanotechnology are likely to extend CMOS technology out to 2015. However, at transistor densities beyond that several problems start to arise. One is the dramatic escalation in the cost of a new fabrication plant to manufacture the chips. These costs must be amortized over the cost of the transistors, keeping them expensive. Second, it becomes increasingly difficult to dissipate the heat caused by the logic devices. Lastly, at such small distances, electrons increasingly tunnel between materials rather than going through the paths programmed for them. As a result of these constraints, any continuation of Moore's Law much beyond 2015 is likely to require the development of one or more new technologies.

Future advances will also bring us closer to a world of free memory, ubiquitous data collection, massive serial processing of data using sophisticated software, and lightening-fast, always-on transmission. What happens when almost all information is theoretically available to everyone all the time?

Cognitive Sciences (Robotics) — Continued advances in computer science combined with a much better understanding of how the human brain works should allow researchers to develop software capable of duplicating and even improving on many aspects of human intelligence. Although progress in Artificial Intelligence has lagged the expectations of many of its strongest proponents, specialized software continues to advance at a steady rate. Expert software now outperforms the best humans in a variety of tasks simply because it has instantaneous access to a vast store of information that it can quickly process. In addition, researchers continue to develop a much better understanding of how individual sections of the brain work to perform specific tasks. As processing power continues to get cheaper, more and more of it will be applied to individual problems.

DOES NANOTECHNOLOGY REPRESENT A DANGER TO SOCIETY?

Few people would doubt that technology has brought great benefits to human society. Even those who are often the most vocal in shunning it are usually quick to adopt those aspects, such as better health and communication, which suit their purposes. In spite of these benefits, society has a love/hate relationship with new advances. This is partially because new technology always creates new economic possibilities, which upset those benefiting from the status quo. Luddites destroyed the first weaving machines because they threatened their existing jobs. The protesters gave little thought to the masses of people who might, for the first time, be able to purchase a second set of clothes at an affordable price. Perhaps deeper is an uneasiness with the uncertainty of where technology might ultimately take us. Is there such as thing as too much progress? Who exactly will benefit? What possible problems lurk and how will we deal with them? What are the social implications? These and other unanswerable questions have often been used as excuses to forego technology's benefits in favor of the comfort of today's problems.

Nanotechnology has generated similar concerns. In perhaps the best known example, Bill Joy, former chief technology officer for Sun Microsystems, wrote an article in which he seriously questioned the wisdom of going forward with current research.[20] Mr. Joy's fears revolved around three possible threats:

- Nanodevices that get out of control. The minuteness of the nanoscale and the vast number of nanoorganisms or devices that are needed to be effective at a macroscale implies a certain loss of control once they are released into the environment. We will have created a lot of them and we will have trouble knowing exactly where they are or what they are doing. Some have expressed the fear that self-replicating nanobots might multiply out of control, eventually consuming all matter and covering the world in a "grey goo." This threat, first raised by Eric Drexler in his book *Engines of Creation* and later the subject of a novel by Michael Crichton has since been widely discredited by most scientists. Beyond the issue that no one now knows how to make self-replicating machines, there are serious questions about how such a process could sustain itself without any clear source of energy. Even Eric Drexler has testified that the grey goo scenario is the wrong issue to focus on.[21]
- The rapid proliferation of the knowledge and equipment needed to create new biological life forms. Mr. Joy is especially concerned that this knowledge intentionally will be used to create and release new pathogens. Unlike nuclear technology, the capacity to create biological weapons of mass destruction requires far less capital investment and is much easier to conceal. This concern is one that will have to be addressed. However, it is very hard to see how society can totally avoid this risk without at the same time giving up on technologies that promise to cure cancer, correct genetic defects, and create new organisms capable of cleaning up toxic chemicals.[22]
- Mr. Joy's final concern was that advances in information technology and artificial intelligence will eventually create an intelligence superior to ours, which may not act solely in our interest. Again Mr. Joy is far more likely to be right about the direction of technology than about its evil effects. The history of technology is that its benefits have vastly outweighed its dangers and that society has eventually found ways to manage even the worst dangers, often using further advances in technology. As with biotechnology, it is hard

to see how society could avoid the possibility of running this danger without at the same time giving up all the benefits that greater automation promises.

Some applications will be harmful and good science is needed to detect and respond to these harms as early as possible. Any broad technology brings both benefits and dangers. As discussed below, certain applications of nanotechnology do present serious environmental and health issues. These applications will have to be monitored and, if the harm outweighs the benefits, curtailed. But such decisions should be made on the basis of sound science, not emotional appeals about the dangers of the unknown. And government policy should reflect the fact that on the whole nanotechnology is expected to bring large net benefits to society and should be encouraged.

Yet, the fear of technology displacing humans runs deep in the human psyche and explains events as diverse as the persecution of Copernicus and Galileo, the Salem Witch Trials, and the continued popularity of Mary Shelley's *Frankenstein* over a century after it was first written. There is also a strong tradition of Luddite opposition to any technology that threatens the existing market of any special interest. Presently, universities, optometrists, realtors, car dealerships, and others are all scrambling to protect themselves from competition enabled by the Internet. The special interests that seek these protections almost always try to justify them as efforts to protect consumers or society.

Any application of technology that causes large costs quickly draws society's attention to it and the costs it imposes provide a strong incentive to correct them. There are therefore reasons to think that, with careful monitoring any product that actually causes severe harm to the environment or health can be removed relatively quickly. Although there are legitimate issues about nanotechnology's effects, any proper discussion of regulation should explicitly acknowledge the danger of letting special interests on either side hijack the process by using legitimate concerns as a pretext for barriers whose main purpose is really to satisfy the interests of narrow groups or to fan unfounded fears. Regulation should also explicitly weigh the risk of inhibiting beneficial uses against the benefit of preventing harmful applications.

Government Policy toward Nanotechnology

We should view government policy in this context. As explained above, nanotechnology is still in its early stages. Many of the most valuable commercial applications are decades away and require continued advances in basic and applied science. As a result, government funding still constitutes a large proportion of total spending on research and development. Within the United States, this spending is guided by the National Nanotechnology Initiative (NNI).[23] The NNI coordinates the policy of 25 government agencies, including 13 that have budgets for nanotechnology research and development.[24] It has set up an infrastructure of over 35 institutions across the country to conduct basic research and facilitate the transfer of technology to the private sector.

The NNI's strategic plan sets out four main goals:[25]

- Maintain a world-class research and development program to exploit the full potential of nanotechnology.

- Facilitate the transfer of nanotechnology into products for economic growth, jobs, and other public benefits.
- Develop educational resources, a skilled workforce, and the supporting infrastructure to advance nanotechnology.
- Support responsible development of nanotechnology.

The NNI is clearly geared toward developing the technology on a broad front, correctly seeing it as the source of tremendous benefits to society. Its mission is not to see whether we should go forward with research and development. It is to go forth boldly, while trying to discover and deal with possible risks.

Despite the fears expressed by Bill Joy, there is relatively little serious debate among policymakers over possible long-term existential threats to mankind. The main topics of discussion are the possible health risks associated with nanoparticles and the need for greater public participation in the development of the technology. Each of these topics is worthy of discussion, but their implications for public policy are much more nuanced than many of their proponents realize. Neither is likely to seriously affect the broad development of these new technologies although they could improve the net benefits that society realizes from them.

A number of concerns have been raised about the effect nanoparticles might have on human health. Precisely because of their small size, there is some fear that they might unintentionally penetrate the normal biological barriers that protect human health. For instance, could a certain particle penetrate human skin, from there cross tissue protecting the brain from foreign chemicals and finally migrate through cell walls to interfere with cell function? Note that in the future, some particles might be specifically designed to do exactly that in order to deliver medication to patients with brain tumors. The concern here is with unintentional exposures. The human body has already evolved defenses against constant exposure to a large variety of nanoparticles, including soot and bacteria. However, in the future many nanoparticles will have novel structures that neither our immune systems nor the environment have ever come into contact with before.

Several animal studies show that certain exposures can lead to health problems, but it is far from clear whether the results have much relevance to the expected exposures humans will face. The central fear is that an engineered particle that is widely used could turn out to be like asbestos or PCBs and have serious long-term health consequences that are recognized only after thousands of people have suffered or large costs have been incurred. In fact, some scientists claim that carbon nanotubes exhibit properties similar to asbestos fibers at the nanoscale.

A recent report by the National Academy of Sciences concluded that: "for now there is very little information and data on, or analysis of, [environmental health and safety] impacts related to nanotechnology" and that "the body of published research addressing the toxicological and environmental effects of engineered nanomaterials is still relatively small."[26] As a result, there has been a widespread call for more greater federal action to address possible health concerns before they arise. Some researchers have called for increasing the government's power to regulate nanoproducts, arguing that existing laws such as the Food, Drug and Cosmetic Act, the Toxic Substances Control Act, and the Occupational Safety and Health Act are inadequate to deal with potential problems.[27] Others have called for significant increases in research on the health effects of nanoparticles and a beeter prioritization of federal spending.[28]

A better understanding of how specific particles affect human health would be enormously valuable. But realizing this will not be as easy as many people would like. First, much knowledge will have to wait for the development of better equipment and facilities capable of measuring quantities and events on such a small scale. The National Academy of Sciences concluded that: "[t]he ability to carry out comprehensive EHS R&D requires that techniques and instrumentation for characterization and measurement be developed that will enable determination of the exact composition of a nanomaterial in a substance or product, as well as the physicochemical properties of specific nanomaterials."[29] Equipment to accurately measure and observe events at the nanoscale is still relatively primitive compared to where it is likely to be in 10-20 years.

Second, spending more money on research does not necessarily mean that the research will be worth the money. Proponents of additional spending are right to point out that, given the relatively small amount currently being spent, the marginal benefits from spending are likely to be high, at least for the next few years. The National Academy of Sciences recommended increasing research on the environmental, health and safety effects of nanotechnology.[30] Although the Academy did not cite a figure, others have called for spending $50 million to $200 million annually.[31] Although this would represent a large increase from the approximately $35 million that the NNI claims to devote to the area now, if properly allocated through peer-reviewed grants by agencies such as the National Science Foundation, such a sum should produce large benefits for several reasons. First, once the results are published they will provide a good base for the private sector to build off of in evaluating the safety of proposed products. Second, the studies should further the knowledge of how engineered nanoproducts interact with biological systems at the cellular level. In addition to making it easier to avoid the production of harmful materials, this general knowledge should make it easier to engineer nanomaterials that accomplish beneficial health purposes. To a large extent, EHS research is a natural complement to efforts to use nanotechnology to combat diseases such as cancer.

But rapid increases in funding do not automatically guarantee rapid increases in results. One important issue is the degree to which agencies should pursue a central list of research priorities. At present, although agencies coordinate through the NNI, each agency retains full control over its own budget decisions and sets its own priorities for research. The National Academy of Sciences concluded that "the NNI is successfully establishing R&D programs with wider impact than could have been expected from separate agency funding without coordination... .The committee believes that federal agencies have been motivated by their participation in NNI activities to establish priorities, coordinate programs, and leverage resources to a degree that has proved very effective."[32]

Although centralization might produce a consistent list of priorities, it does not always produce the best one. If centralization might steer funding toward important areas that the agencies might normally view as being outside their narrow areas of concern, it might also fail to fund some areas of research that are central to an agency's mission. Centralized priorities are only as good as the process used to establish and implement them. Given that the NNI is presently based purely on collaboration, the best alternative would probably be to give the NNI a significant portion of independent budgetary authority that it could use to fund research in areas that fall in between or overlap the interests of the separate agencies. An independent budget would also give the NNI greater weight in guiding the agencies toward consistent progress on developing a coordinated nanotechnology policy but still leave the latter free to pursue their own mandates.

It is also very clear that health research must be better coordinated with the private sector and government agencies at the state and international levels. Experiments done in one part of the world have immediate relevance to all other areas and there is great benefit in avoiding duplication and spreading research findings widely. The benefits of coordinating research among domestic and international laboratories are significant.

A final issue concerns the obligations that private companies should face in ensuring the safety of the products they sell in the market. In many cases, such as cosmetics, these products face very little regulatory scrutiny prior to reaching consumers. The combined lack of testing and oversight has led at least one organization to call for a moratorium on the further commercial release of personal care products that contain engineered nanomaterials and the withdrawal of products currently on the market.[33] The general issue of risk is discussed in greater detail below. But one legitimate concern is a lack of information on the amount and type of testing that testing companies perform in order to ensure that their products are safe. Under current law, companies are not required to disclose the results of any safety testing and many companies consider such research proprietary.

The debate on the safety of using nanotechnology would be improved if three changes were made governing the use of nanotechnology in products. First, the use of nanotechnology should be clearly labeled on products so that consumers can make an informed choice about whether to use a particular product. At present, manufacturers are split on the marketing value of nanotechnology. Some tout it in their advertising even if their product does not technically contain nanoparticles, on the theory that consumers are attracted to new technology. Others, fearing a consumer backlash if consumers develop a negative view of nanotechnology, omit any mention of the word. Clearer labeling of exactly what ingredients are used and of the particle size would give consumers accurate information and reduce the possibility of a sudden backlash if there is a problem with one or more specific products. Consumers ought to have the ability to make independent judgments about whether to purchase products with nanoparticles.

Second, private companies should be required to disclose to the Food and Drug Administration the results of any safety testing that they conduct and the FDA should immediately publicize any results that show a clear negative health effect. Companies would then probably find it in their interest to publicize neutral or positive findings. Disclosure of test results does have important strategic implications for companies that compete for market share. But, since most safety testing will be done by the private sector, members of the public should have the right to see what steps companies are taking to protect their health. This would also ensure that the debate over safety occurs in public with full information. While this might subject companies to some discomfort in the short-term, it will make it much more difficult for opponents of the technology to use public distrust to exploit any negative stories. Congress could encourage additional safety testing by making it easier for companies to collaborate on precompetitive research into the environmental, health and safety impact of nanomaterials.

Additional efforts to identify the environmental, health and safety risks of nanoparticles will bring clear benefits. But the need to conduct these studies should not be used to prevent the introduction of new products. Science and technology have always involved a leap into the unknown, bringing with it an assumption of unforeseen risks. Opponents of technology can always point to examples of innovation gone bad such as asbestosis, DDT, PCBs. But their analysis of this risk omits three important facts. First, each of these products brought with them significant benefits which, at least for a while, could not be duplicated by other products. Indeed DDT has recently been reapproved for limited use to combat malaria. Second, even if the total cost of these products outweighed their benefits, the former were unnecessarily increased by a lack of full disclosure about research into their health effects. That is why an open debate about EHS testing is so important. It allows society to improve the cost/benefit equation of any given product. Third, and by far the most important, any testing policy that significantly delayed the use of these products might have also delayed the use of thousands of other products that did not prove to pose significant health risks. This would have had major impacts on economic growth and consumer welfare. Any policy that tries to stop harmful products from entering the market must try to do so without significantly delaying the vast majority of products that bring net benefits.

One environmental group has made clear its position on nanotechnology. It calls for a moratorium on all products containing nanomaterials. In their words: "We believe that ethical concerns and the likely far-reaching socio-economic impacts of nanotechnology, must be addressed alongside concerns over nanotoxicity before the commercialization of nanotechnology proceeds." One of the many criteria that they require to be met before nanomaterials can be commercially released is that "safety assessments are based on the precautionary principle and the onus is on proponents to prove safety, rather than relying on an assumption of safety.[34]

Rather than being an impartial look at the possible health risks of using nanoparticles in cosmetic products, the report is a biased swipe against a broad category of consumer products. In the case of sunscreens there is no discussion of the possible benefits that might occur if more people either use more sunscreen or find its use more enjoyable because standard ingredients such as zinc oxide and titanium dioxide appear clear rather than white at the nanoscale. This sort of possible benefit is simply assumed not to be important. Having established that the benefits are zero, the report then looks at the risks. Here its discussion is similarly one-sided. Although it cites the 2004 report of the Royal Society at least 14 times, it never mentions the report's discussion of the regulatory approval for titanium dioxide. Nor does it point out that the Royal Society specifically found that a moratorium on nanotechnology was not justified.

Of course, it is unclear how the standard advocated by Friends of the Earth could ever be met. Few products come with an absolute guarantee of safety for all portions of the population. Under this standard it would not be enough if a product's cost/benefit ratio was positive or even very high. The question would be whether the product imposed any risks to society at all. And if there was even the possibility of a risk (and there would almost always be at least the possibility) then the product could be rejected. Proponents of growth should always remember that there is a certain section of the population that argues against the introduction of peanuts because exposure can be deadly to those with a strong allergic reaction to them. Similar arguments can and will be made against nanotechnologies even when an impartial cost/benefit evaluation shows that the technology will probably bring net benefits to society.

The requirement to address "the far-reaching socio-economic impacts" also imposes an almost insuperable barrier. First, many of these impacts are unknowable because they depend on a variety of other events in the future. Widely used technologies do not impact society as single items. They combine to constitute a web of technology that changes the entire social system. It is usually meaningless to pick out one possible application of the technology and evaluate it apart from all the complementary and competing technologies that affect its impact on society. Second, many of the most significant impacts will occur because nanotechnology brings with it large benefits and therefore becomes infused into a wide variety of products in many industries. Most of those who are negatively affected by it will be so because the technology opens up new production, distribution and profit opportunities. They will quickly use arguments against the technology to seek competitive protection.

There is a widespread desire to avoid repeating the mistakes of biotechnology, a technology whose advance has been substantially slowed by political opposition that has little scientific basis. But it is not really clear what the mistakes of biotechnology are. No human deaths can be attributed to genetically modified organisms. Nor has any product of biotechnology ever resulted in significant environmental harm. The potential health and environmental benefits of biocrops in the form of reduced use of pesticides, fertilizer, and fuel and improved vitamin delivery are totally discounted in favor of vague warnings against Frankenfood. One might wish that companies like Monsanto had been more open about their research and intentions, but this research surely would have been used against them by environmental groups who intentionally distort the debate by exaggerating any dangers and denying any benefits. It is far from certain that better studies and more open debate would have produced a more reasoned policy.

Much of the reaction against nanotechnology is based solely on the fact that even if it has benefits, these benefits will change society in substantial ways. This is why opponents often mention the need to look at "socio-economic effects". Similar arguments are being used today against the expansion of the Internet. Realtors have argued that home searches done over the Internet are not really the same as those done by a licensed professional and that the industry therefore should not have to open up its listing services to discount brokers. Optometrists have argued that contact lenses purchased over the Internet are not really as safe as those that they sell and that therefore they should be allowed to write prescriptions for brands that promise not to make their products available to Web stores. Of course, in neither of these arguments is there room for the consumer to determine what actually does or does not benefit him. Rather, the strategy is for the incumbents to make the decision for the individual. Had the development of the World Wide Web waited for a full understanding of its "socio-economic effects" it would probably not exist today.

In this context it is worth discussing what role the public should play in guiding the progress of nanotechnology. The NNI has defined seven Program Component Areas under which it groups related projects and activities. One of the Program Component Areas is devoted to the societal dimensions of nanotechnology. Within this category the NNI intends to foster the following activities:[35]

- Research on the environmental, health and safety impacts of nanotechnology;
- Educational activities including the development for materials for schools, technical training and public outreach; and
- Research on the broad implications of nanotechnology, including social, economic, ethical, and legal implications.

This implies an intent to educate the public about the benefits (or costs) and progress of nanotechnology. Proponents of public education and EHS research frequently point to biotechnology as a lesson of why such efforts are needed. The belief is that after a very promising start, progress in biotechnology has been slowed, and in some cases even halted, due to a broad public reaction that is fueled by:

- Public health scares, although in most cases these had nothing to do with biotechnology. A good example is the damage caused by mad cow disease in England and the rest of Europe. Government delay and deception in dealing with this issue led to a significant decline in the public's confidence about the government's commitment to safety regulation, which opponents of biotechnology exploited.
- The lack of outreach and openness on the part of biotechnology companies such as Monsanto. These companies took the lack of public opposition for granted and did not respond rapidly to questions about the safety or economic benefits of their products.
- Lack of general public education about either the science or the economic benefits of genetically modified crops. This lack of knowledge provided little perspective with which to judge conflicting health claims. Since consumers did not know of any benefits biotechnology might bring, they had little reason to miss them.
- A determined opposition by some environmental groups that were adamantly opposed to the use of genetically modified crops under any conditions, regardless of the science. These groups engaged in a determined campaign to convince the public that biotechnology represents a grave threat to the public health and the environment. They took legitimate questions and expanded them into worst case scenarios and then made those scenarios seem like a certainty if a complete ban was not enforced. They often used violence to enforce their beliefs and gain publicity for their cause.

It is hard to argue against public education. The public should have a voice in how public money is spent, and it should be an informed voice. Even within the NNI budget, allocations between theoretical research, medical applications, and EHS studies are subjects of legitimate debate.

But it is important to have a realistic view of what public engagement can accomplish. As we go forward, an increasing proportion of investment in nanotechnology will come from the private sector. As a result, government will gradually lose much its ability to shape the direction of in which the technology advances. Decisions will increasingly be made by a decentralized collection of international businesses, universities, consumers and investors. Any attempt to subject these decisions to a collective decision process in order to manage broad "socioeconomic effects" is almost certain to do far more harm than good. But because the harm from overly stringent regulation will come mainly in the form of future beneficial technology that will be delayed or stopped altogether, it may not be immediately apparent. Government should, however, be involved in monitoring technological developments, identifying any specific environmental risks, holding manufacturers responsible for any harm that their products do cause, and, where appropriate, implementing carefully designed regulatory systems justified by careful cost/benefit analysis.

Nanotechnology must be allowed to proceed as other transforming technologies such as chemistry, steam power, and electricity have done. It must proceed at its own pace and in its own direction. Better dialogue and research can help society deal with specific problems as they become apparent. It can also address the inevitable economic dislocation that will affect specific markets. But policymakers should not fool themselves into thinking that a collective political process can guide the future any better than the market can. Regulations need to be based on clear cost/benefit calculations supported by scientific evidence. And regulations to address specific identified risks should not delay the advancement of a broad range of products that will surely bring large social and economic benefits.

The world in which our children live will surely be a different one. Whether it is a better one is largely up to them to decide. Continued technological advancement, including on the nanoscale, will not automatically make the world any fairer or safer, but it will increase the resources available to those who want to ensure that it is.

Joseph Kennedy
Senior Economist

End Notes

[1] *Future Shock,* Amereon Ltd. (1970).
[2] *Small Wonders, Endless Frontiers: A Review of the National Nanotechnology Initiative*, National Research Council, Washington D.C., 2002, p. 5.
[3] *The National Nanotechnology Initiative at Five Years: Assessment and Recommendations of the National Nanotechnology Advisory Panel*, President's Council of Advisors on Science and Technology, Washington D.C., May 2005, p. 7.
[4] *Nanoscience and Nanotechnologies: Opportunities and Uncertainties*, Royal Society and The Royal Academy of Engineering, UK, July 2004, p. 5.
[5] Markoff, John, "A Chip That Can Move Data at the Speed of Laser Light," *New York Times*, September, 18, 2006, p. C1.
[6] "Powerful Batteries That Assemble Themselves." *Technology Review*, available at http//:www.technologyreview.com/printer _friendly _article.aspx?id=1 7553.
[7] Richard Booker and Earl Boysen, *Nanotechnology for Dummies*, Wiley Publishing Inc., (2005) pp. 142- 44.
[8] The National Nanotechnology Initiative Strategic Plan, National Science and Technology Council, Washington D.C., December 2004, p. i.
[9] Mark Ratner and Daniel Ratner, *Nanotechnology: A Gentle Introduction to the Next Big Idea,* Prentice Hall (2003) p. 13.
[10] M.C. Roco, "Nanoscale Science and Engineering: Unifying and Transforming Tools" *AIChE Journal* Vol. 50, No. 5, pp. 895-6. Until recently, Dr. Roco chaired the U.S. National Science Technology Council's Subcommittee on Nanoscale Science, Engineering and Technology.
[11] M.C. Roco, "International Perspective on Government Nanotechnology Funding in 2005," *Journal of Nanoparticle Research*, Vol. 7, No. 6, p. 707.
[12] M.C. Roco, "Nanoscale Science and Engineering: Unifying and Transforming Tools" *AIChE Journal* Vol. 50, No. 5, p. 896.
[13] *Id.*
[14] Ray Kurzweil, *The Singularity is Near: When Humans Transcend Biology*, Viking Press (2005).
[15] *See,* Microsoft Corporation, *Toward 2020 Science,* available at http://research.microsoft.com/towards2020science/downloads/T2020S_ReportA4.pdf
[16] National Academy of Sciences, *Small Wonders, Endless Frontiers,* Washington D.C., 2002, pp. 3 6-45.
[17] Ray Kurzweil, *The Singularity is Near: When Humans Transcend Biology*, Viking Press (2005), pp. 205- 98.
[18] Joel Garreau, *Radical Evolution: The Promise and Peril of Enhancing Our Minds, Our Bodies – and What it Means to Be Human,* Doubleday (2005).
[19] Mihail C. Roco, "The Emergence and Policy Implications of Converging New Technologies," In William Sims Bainbridge and Mihail C. Roco (Eds.), *Managing Nano-Bio-Info-Cogno Innovations: Converging Technologies in Society*, Springer (2006), pp. 8-22.
[20] Bill Joy, "The Future Doesn't Need Us" *Wired,* April 2000.
[21] The Royal Society and Royal Academy of Engineering, *Nanoscience and Nanotechnologies: Opportunities and Uncertainties,* RS Policy Document 19/04, July 2004, p. 109.
[22] *See also*, Mark Williams, "The Knowledge," *Technology Review*, (March/April 2006), p. 44.

[23] Significant legislation governing the NNI includes the 21st Century Nanotechnology Research and Development Act (P.L. 108-153). Interagency coordination is managed by the Nanoscale Science, Engineering, and Technology Subcommittee within the National Science and Technology Council.

[24] For a good description of the NNI, see, National Science and Technology Council, *The National Nanotechnology Initiative: Supplement to the President's 2007 Budget*, Washington D.C., July 2006. Available at http://www.nano.gov/NNI_07Budget.pdf.

[25] National Science and Technology Council, *The National Nanotechnology Initiative, Strategic Plan,* Washington D.C., December 2004. Available at: http://www.nano.gov/NNI_Strategic_Plan_2004.pdf.

[26] *A Matter of Size: Triennial Review of the National Nanotechnology Initiative*, National Academy of Sciences, Washington D.C. 2006, p. 78.

[27] *See*, J. Clarence Davies, *Managing the Effects of Nanotechnology,* Woodrow Wilson International Center for Scholars, Project on Emerging Nanotechnologies, Washington D.C. Available at: http://nan otechproject.org/index.php?s=reports.

[28] See, Andrew D. Maynard, *Nanotechnology: A Research Strategy for Addressing Risk,* Woodrow Wilson International Center for Scholars, Project on Emerging Nanotechnologies, Washington D.C., July 2006. Available at: http://nanotechproject.org/index.php?s=reports.

[29] *A Matter of Size: Triennial Review of the National Nanotechnology Initiative*, National Academy of Sciences, 2006, p. 80.

[30] *Id.* p. 92.

[31] See, Andrew D. Maynard, *Nanotechnology: A Research Strategy for Addressing Risk,* Woodrow Wilson International Center for Scholars, Project on Emerging Nanotechnologies, July 2006; Testimony of Matthew M. Nordon, President, Lux Research Inc., September 21, 2006, U.S. House of Representatives Committee on Science.

[32] *A Matter of Size: Triennial Review of the National Nanotechnology Initiative*, National Academy of Sciences, 2006, p. 6.

[33] *Nanomaterials, Sunscreens, and Cosmetics: Small Ingredients Big Risks,* Friends of the Earth, May 2006.

[34] *Id.*, p. 17.

[35] National Science and Technology Council, *The National Nanotechnology Initiative, Strategic Plan,* Washington D.C., December 2004. Available at: http://www.nano.gov/NNI_Strategic_Plan_2004.pdf.

CHAPTER SOURCES

The following chapters have been previously published:

Chapter 1 – This is an edited, excerpted and augmented edition of a United States Congressional Research Service publication, Report Order Code RL34511, dated June 2, 2010.

Chapter 2 – This is an edited, excerpted and augmented edition of a United States Congressional Research Service publication, Report Order Code RL34401, dated May 13, 2010.

Chapter 3 – This is an edited, excerpted and augmented edition of a United States National Technology Initiative Publication, dated February 2010.

Chapter 4 – This is an edited, excerpted and augmented edition of a United States President's Council of Advisors on Science and Technology publication, dated March 12, 2010.

Chapter 5 – This is an edited, excerpted and augmented edition of a United States Congress Joint Economic Committee Publication, dated March 2007.

INDEX

C

D

E

J

K

L

M

Q

R